Geeks, Genes, and
the Evolution of Asperger Syndrome

Frontispiece. Self-portrait Eve drew when she was eleven years old.

Geeks, Genes, and the Evolution of Asperger Syndrome

Dean Falk and Eve Penelope Schofield

University of New Mexico Press ■ Albuquerque

Library of Congress Cataloging-in-Publication Data
Names: Falk, Dean, author. | Schofield, Eve Penelope, 1991– author.
Title: Geeks, Genes, and the Evolution of Asperger Syndrome / Dean Falk and Eve Penelope
    Schofield.
Description: Albuquerque: University of New Mexico Press, 2018. | Includes bibliographical
    references and index. |
Identifiers: LCCN 2017026650 (print) | LCCN 2017047256 (ebook) | ISBN 9780826356932
    (E-book) | ISBN 9780826356925 (pbk.: alk. paper)
Subjects: LCSH: Schofield, Eve Penelope, 1991—Health. | Asperger's syndrome—Patients—
    Biography. | Asperger's syndrome—Research.
Classification: LCC RC553.A88 (ebook) | LCC RC553.A88 F35 2018 (print) | DDC 616.85/8832—
    dc23
LC record available at https://lccn.loc.gov/2017026650

Cover photograph courtesy of Sarah Nichols
Cover designed by Lisa Tremaine
Interior designed by Felicia Cedillos
Composed in Minion Pro 10.25/14.25

Dean: In loving memory of Alfred and Jane Falk.

Eve: To all who were at the Kingsweston Centre at Oasis Academy, Brightstowe, for helping me to grow into my skin.

CONTENTS

If you met my twenty-six-year-old granddaughter and coauthor of this book, Eve, you would probably be struck by her large vocabulary and odd mannerisms. Perhaps you wouldn't be able to put your finger on it, but if you were familiar with Asperger syndrome (AS) it would not take long for you to wonder if Eve has it, which she does. Like other people with AS, Eve had to be taught how to make eye contact and engage in small talk. As a young adult, she still has difficulty grasping the nuances of other people's tone of voice (and modulating her own) and understanding the thoughts and desires of others (an ability that psychologists call "theory of mind"). Despite these challenges, Eve read precociously as a child, and I am proud to say she recently earned a bachelor of art's degree in creative writing and publishing from Bath Spa University in England. As you will see, Eve has avid interests in Japan, anime, and fantasy-based fiction, interests she will cheerfully share, more or less nonstop, with anyone who will listen.

I am a paleoanthropologist who has spent the last four decades researching the evolution of the human brain and cognition, and I have a longstanding interest in how and why language, music, art, and science emerged in our ancestors. I have been privileged to study materials ranging from the fossilized remnants of the brains of our prehistoric relatives to the cerebral cortex of Albert Einstein. The research, clinical politics, and controversy that continue to engulf AS have been on my radar since Eve was first diagnosed with it when she was nine years old.

As is well known, Asperger's Disorder was folded into Autism Spectrum Disorder along with certain other developmental disorders in the most recent revision of the *Diagnostic and Statistical Manual of Mental Disorders* published by the American Psychiatric Association.[1] Consequently, many clinicians and researchers no longer recognize Asperger's in its own right. This change has polarized the autism community, as discussed in the introduction.

For now, suffice it to say that if the American Psychiatric Association were a society of evolutionary biologists instead of psychiatrists, a parallel revision might be to announce that apes and people are part of a broad spectrum of primates (which is true) and to banish recognition of individual primate species (no more *Homo sapiens, Gorilla gorilla,* and so on). If followed, this rule would make it exceedingly difficult to learn meaningful details about the hundreds of varieties of primates that grace our planet today. However, as scholars have known since the 1735 publication of the system that is still used to classify all forms of life,[2] there is a time for "lumping" individuals into larger groups (such as primates) and a time for "splitting" them into narrower categories (such as monkeys, apes, and humans). In other words, there is no one "right" way to classify organisms into groups. The same may be said for autism. Sometimes it is useful to think of it as consisting of one large spectrum, for example, when considering its global prevalence, which we will do later in the book. At other times, it makes more sense to recognize that the broad spectrum consists of different kinds of autism.

For example, the estimated 50 percent of autists who remain mute throughout their lives can be distinguished from those who eventually develop language. The latter group may be divided further into individuals who were delayed in the development of language compared to typically developed children (often described as having high-functioning autism, or HFA) and autists who were not delayed in acquiring language (identified in the now-discarded *DSM-IV* as having Asperger's Disorder). Although all these groups are worthy of investigation in and of themselves, we focus primarily on AS and HFA, not because other kinds of autism aren't important and interesting but because AS and, to some extent, HFA are well suited for evolutionary analyses.

The vocabulary one uses for autism is important because, among other reasons, the label a child receives will likely influence his or her future. With respect to the terms "low-functioning autism" and "high-functioning autism," for example, science writer Nicholette Zeliadt observes that "either label can be limiting: It might prevent one child from participating in activities she wants to do, or exclude another from getting the services he needs."[3] As Zeliadt details, autism research has been plagued by the problem of precisely defining these terms and determining to whom they should apply. This may account, in part, for an increasing backlash against using these labels among some autism researchers and parents. Zeliadt notes, however, that "abandoning labels altogether isn't an alternative." Some would like scientists to replace the terms "low-functioning" and "high-functioning" with "high-support" and "low-support" (that is, the support needed by the autistic child).[4] As noted, this book synthesizes information from many scientific studies that have focused on AS and HFA. We use the terms and definitions (including for HFA) employed by the authors of the various studies discussed in this book, although we are sympathetic with the viewpoint that many individuals who have been diagnosed with low-functioning autism (who are not a focus of this book) may, in fact, function well in day-to-day aspects of their lives.

A word about some of the other terms used in this book. We prefer "Asperger syndrome" to the *DSM-IV* label of "Asperger's Disorder" (but retain the latter when the context calls for it), partly because Hans Asperger himself preferred "syndrome," as indicated when he wrote about "Asperger syndrome, as it is now known."[5] Further, we hesitate to characterize AS as a disorder because, as this book explains, its essence is best understood as a natural byproduct of positive past and present evolutionary dynamics. We also follow others in using "autist"[6] and/or "autistics"[7] interchangeably as a general designation for people with autism. Similarly, "neurotypical" (coined by the first autistic-run organization, the Autism Network International)[8] and "typically developed" are labels for unaffected people who frequently comprise the comparison (control) groups in autism studies.

Because "Aspies" is more concise than phrases such as "persons with Asperger syndrome" or "individuals with AS," and because Eve prefers it, "Aspie(s)" is used widely, but not exclusively, throughout the book. By adopting this term, we follow autism advocate Judy Singer, who used it when she noted in 1999, "The people who band together under this category prefer to name their condition as AS, and themselves as autistics, and sometimes, comfortably, as 'aspies,' to distinguish themselves from those they have dubbed the 'NTs'—Neurotypicals."[9] Aspie author Liane Holliday Willey was also among the first to use "Aspie,"[10] by which she "meant it to take away the stigma associated with syndrome. I do not want a syndrome, nor do I want to live my life under the umbrella of a man's name. But whatever my reason in choosing to refer to myself as an aspie, I never meant for it to imply anyone in my Asperger Syndrome community was or is crazy, insignificant, separate and unequal, or anything remotely negative."[11] Similarly, in this book "Aspies," "autistics," "autists," "high-functioning autism," "neurotypical," and "typically developed" are intended to be respectful. So is our use of "geeks" to identify highly talented technical-minded individuals.

In each chapter, I discuss the latest findings related to a specific facet of AS (and, where possible, HFA), such as the cognitive, genetic, or neurological underpinnings of these conditions. Put simply, my goal is to understand how AS fits within the evolutionary-developmental (evo-devo) framework that caused humans to emerge as the most cognitively advanced species on the planet. Eve, on the other hand, writes about experiences she has had that are relevant to some aspect of each chapter's scientific content. (Unlike her grandmother, Eve uses British spellings in her parts of the book.) Because her opinions are those of only one person, her parts of the book are obviously not meant to represent all people with AS. They do, however, give a feel for what it is like to be an Aspie. In the last chapter, Eve offers suggestions about how parents of children with AS and HFA can help them have a better future.

Discussion of the latest neurophysiological or genetic findings requires going into some detail, which means that parts of this book are, of necessity, somewhat technical. Interested readers will find still more information in the notes, the glossary, and an extensive list of references. Others may wish to skim the technical discussions. At whatever level you choose to read this book, we hope that Aspies, other autists, parents, teachers, clinicians, psychologists, psychiatrists, other health-care providers, autism researchers, evolutionary biologists, paleoanthropologists, and people who simply enjoy reading about science will find thinking about AS and HFA from an evolutionary perspective thought provoking—and maybe even fun.

We are immensely grateful to the University of New Mexico Press and, especially, to our editor, John Byram, for taking on this book project. Dr. Michael Margolius is warmly acknowledged for diligently providing references and news about autism as our writing progressed. Professor Chris Gunter is thanked for providing constructive suggestions on an earlier draft. We thank Karin Kaufman for careful copyediting and Tony Archer, director of creative services at Florida State University, for help with illustrations. Michael Brown, president at the School for Advanced Research, and Rochelle Marrinan, chair of the Department of Anthropology at Florida State University, are acknowledged for their generous support.

We are also grateful to the following people who have helped move this book forward: James Ayers, Amanda Baxter, Colette Berbesque, James Brooks, David Cohen, Alyssa Crittenden, Margaret Dancy, Arthur Davis, Doug Dearden, Pete Ellis, Karen Foulke, Russell Dean Greaves, George (Jorge) Gumerman, Sheila Gumerman, Elizabeth Hadas, Pier Jaarsma, Janet Kistner, Daniel Lieberman, Jack Meinhardt, Deirdre Mullane, Sarah Nichols, Michael Peters, Fred Prior, Vanessa Panetta Reinhardt, Karen Rosenberg, Jean Schaumberg, Aidan Schofield, Judith Schofield, Lena Dean Schofield, Suzanne Staszak-Silva, Kenneth Stilwell, Laurel Trainor, Wenda Trevathan, Garret P. Vreeland, Sandra Vreeland, and Katherine White.

Finally, we thank Joel Yohalem for his thoughtful feedback on all drafts of the manuscript, patience, and support—Joel, always Joel.

AS	Asperger syndrome
ASD	Autism Spectrum Disorder
AQ	Autistic quotient
CDC	Centers for Disease Control and Prevention
CNV	Copy-number variation
DSM-IV	*Diagnostic and Statistical Manual of Mental Disorders,* 4th edition
DSM-5	*Diagnostic and Statistical Manual of Mental Disorders,* 5th edition
evo-devo	Evolutionary developmental
EQ	Empathizing quotient
fMRI	Functional magnetic resonance imaging
HFA	High-functioning autism
MRI	Magnetic resonance imaging
SQ	Systemizing quotient
TD	Typically developed
ToM	Theory of mind
TPJ	Temporo-parietal junction

I n 1944, Austrian pediatrician Hans Asperger described a clinical condition that eventually became known as Asperger syndrome.[1] Because his young patients were intensely focused on their inner worlds, Asperger described them as "autistic," which literally means "state of being self-absorbed." A year earlier, psychiatrist Leo Kanner used the same label for another group of children, who were later recognized as having "classic" autism.[2] Most of the patients in both groups were boys who were socially inept and insensitive to the thoughts and feelings of others. They avoided eye contact and had flattened or otherwise strange tones of voice. Instead of playing imaginatively with others, they engaged in repetitive solitary activities, like forming patterned rows with toys or organizing objects into stacks. The youngsters hated change and became terribly upset when their routines were disrupted.

After he learned of Kanner's research, Asperger observed remarkable differences between the two groups of children. Referring to himself in the third person, he wrote:

> Asperger's typical cases are very intelligent children with extraordinary originality of thought and spontaneity of activity. . . . Their thinking, too, seems unusual in that it is endowed with special abilities in the areas of logic and abstraction. . . . A further important difference . . . is that Asperger children, very early, even before they walk, develop highly grammatical speech and they may be uncommonly apt at using expressions coined spontaneously. . . . However, the children with Kanner's syndrome generally avoid communication, consequently they do not develop speech or develop it very late. . . . The Asperger type of child . . . may continue with his special subject with undiminished vigour and with originality and may in the end find his way into an unusual career, perhaps into highly specialised scientific work, maybe with an ability

bordering on genius. . . . Indeed, it seems that for success in science or art a dash of autism is essential. For success the necessary ingredient may be an ability to turn away from the everyday world, from the simply practical, an ability to re-think a subject with originality so as to create in new, untrodden ways, with all abilities canalised into the one specialty.[3]

Half a century after Asperger published his findings, the American Psychiatric Association named Asperger's Disorder as a pervasive developmental disorder in its 1994 edition of the *Diagnostic and Statistical Manual of Mental Disorders (DSM-IV)*.[4] The diagnostic criteria for Asperger's Disorder included significantly impaired social interactions, repetitive patterns of behavior, and highly restricted interests and activities. *DSM-IV* also specified that affected individuals experienced no clinically significant general delay in language or cognitive development. Above all, it was this specification that set Asperger's Disorder apart from another condition listed in *DSM-IV*, Autistic Disorder. The formal recognition of Asperger's Disorder was extremely important because a diagnosis based on the highly respected *Manual* qualified children for medical and educational services. Eve was among the youngsters who benefited. Having been diagnosed with Asperger's at age nine, she became eligible to receive treatments that helped her adjust socially and become comfortable in her Aspie skin.

As is well known, the proverbial rug was pulled from under the Asperger community in 2013, when the American Psychiatric Association folded Asperger's Disorder, Autistic Disorder, and several other conditions into a single category called Autism Spectrum Disorder (ASD) in the fifth (and latest) edition of its *Manual (DSM-5)*.[5] What this meant for undiagnosed individuals who would have met the criteria for Asperger's Disorder using *DSM-IV* guidelines was that in the future they might or might not satisfy the comparatively restrictive criteria for ASD in *DSM-5* and, thus, might or might not become eligible for needed services and treatments.[6]

But it's not just access to services that was, and is, at stake. The public attaches a stigma to autism but is more neutral and, sometimes, even positive about Asperger's. One critic of *DSM-5* raised the specter that "children and adults who are shy and timid, who have quirky interests like train schedules and baseball statistics, and who have trouble relating to their peers—but who have no language-acquisition problems—[will be] placed on the autism spectrum," which "often does a real disservice" and can lead to lower self-esteem and fewer job opportunities.[7] Clearly, it is not always easy to draw the boundary between the high-functioning end of the autism spectrum and social oddity.[8]

The motivation for merging several previously recognized pervasive developmental conditions into one big spectrum in *DSM-5* was to establish "a new, more accurate, and medically and scientifically useful way of diagnosing individuals with autism-related disorders." This was premised, in part, on the assumption that "symptoms of people with ASD will fall on a continuum, with some individuals showing mild symptoms and others having much more severe symptoms."[9] A belief that Asperger's Disorder melded seamlessly at the mild end of the spectrum with high-functioning autism was implicit in this rationale, although HFA was not (and is not) a formally recognized form of autism. Instead, high-functioning autists who did not meet the criteria for Asperger's Disorder were likely to be diagnosed with Autistic Disorder in *DSM-IV*, although the criteria for distinguishing the two disorders were somewhat fuzzy.[10]

Although there is no doubt that people with Asperger's are both autistic and high functioning as, indeed, Asperger himself describes in the passage quoted above, the widely held belief that AS and HFA are for all practical purposes identical requires careful scrutiny. Because of this assumption, a *huge* number of recent studies compare groups of individuals that have *either* Asperger's Disorder or HFA with neurotypical control groups. Fortunately, some investigations still compare groups comprised only of Aspies with groups of neurotypicals.[11] In order to glean as much information as possible about the evolutionary emergence of HFA and AS, this book incorporates findings from both types of studies.

Whether or not clinicians identify autists as low or high functioning is usually determined by their performances on full-scale IQ tests. A recent analysis of fifty-two separate studies showed that people with HFA were extremely variable when it came to full-scale IQs. Their scores fell across ranges of IQs described as "above . . . intellectual disability" (70–79), low average (80–89), average (90–109), or high average (110–119).[12] Significantly, the full-scale IQ ranges for AS and HFA overlapped to some degree, but not entirely, consistent with the authors' conclusion that AS and HFA represent different subtypes of autism.[13] That said, it is important to emphasize that, as is true for Aspies, some autists with HFA are remarkably intelligent—animal scientist Temple Grandin comes to mind, for example.[14] Although there are high-IQ (even gifted) individuals in both groups, they are still distinguished by certain cognitive differences.[15]

And make no mistake, Aspies constitute an identifiable subgroup of autists. Ironically and as mentioned above, the most important difference between people with Asperger's Disorder and those with HFA was spelled out in the old *DSM-IV*: Aspies have no significant delay in their development of language, which according to *DSM-IV* meant they used single words by two years of age and communicative phrases by age three. Other autistic children, however, are frequently delayed in language development or do not develop it at all. (In fact, an estimated 50 percent of all autists remain mute.)[16] For the purposes of this book, HFA is, thus, an informal designation for autistic individuals who were delayed in their development of language but otherwise unimpaired in terms of general intelligence. Clearly, timely acquisition of language is the main thing that sets Aspies apart from other high-functioning autists.

But it is not just language development that distinguishes Aspies, on average, from people with HFA. More than a hundred studies that compared individuals in the two groups revealed significant differences in their core clinical features, motor and sensory functions, neuropsychological and neurocognitive traits, brain structures, genetic underpinnings, co-morbidities, and treatment outcomes.[17] Aspies' performances tend to be higher on tests of vocabulary, information, and arithmetic.[18] One study that investigated whether HFA and AS are distinct disorders or simply the same one with different degrees of severity found that Aspies had "strengths on verbally mediated skills as well as weaknesses on visual-motor coordination and graphomotor ability, whereas the children with HFA exhibited a profile with deficits on tasks calling upon verbal comprehension and good performances on tasks requiring visuo-spatial skills."[19] Taken together, these findings are striking, as indicated by the title of a recent review: "Asperger's Disorder Will Be Back."[20] In fact, one could argue that AS never really went away, as evidenced by the voluminous amount of new research that continues to explore its biological, neurological, and genetic underpinnings.[21]

An underappreciated distinction, but one recognized by Hans Asperger (see the above quotation), is that Aspies develop not only timely speech but also grammatical speech, which appears to be impaired in other forms of autism, including HFA.[22] Grammatical speech is based on acquired language-specific rules for forming words (e.g., making them plural or changing their tense) and organizing them into phrases and sentences, which facilitates communication of an endless variety of ideas. It depends on precise anatomical acrobatics (involving the vocal cords, tongue, teeth and lips) that produce, segment, assemble, and utter meaningful streams of vibrating air. Complex grammatical speech is something that only humans can do (and comprehend), and it depends on highly evolved brains that are wired for certain kinds of exquisitely rapid sequential, analytical, and logical processing. Other human endeavors such as creativity in the arts and sciences also make use of this advanced neurological machinery and are just the kinds of abilities for which, as Hans Asperger observed, "a dash of autism is essential."

There is no doubt that Aspies excel at the types of linguistic, systematic, computational, analytical, and creative thinking that evolved in our prehistoric ancestors. This is one reason I argue that AS emerged relatively recently as a natural byproduct of three advanced evolutionary-developmental (evo-devo) trends that were sparked by evolutionary pressures on infants millions of years ago. As detailed in chapter 1, these trends ultimately paved the way for the emergence of language and other uniquely human intellectual spinoffs.[23] A less intellectual but equally fascinating spinoff of prehistoric evolution is reflected in the development of unusual sensory systems in people with HFA and AS, which is the topic of chapter 2.

Chapter 3 discusses cognition and the brain in AS and HFA. Part of the discussion is grounded in details about the evolution of different specializations in the left and right halves (hemispheres) of the human cerebral cortex. The linguistic and analytical skills of most people depend largely on the left sides of their brains, which became rewired during prehistory. This did not happen in isolation, however. The right sides of ancestral brains simultaneously became reorganized to process information in a more intuitive, holistic, musical, and visuospatial manner than occurs in the sequentially driven left hemisphere. Although the two sides of the brain typically connect with one another to process information in an astonishingly rapid and coordinated manner, this is not always the case for autists.

An idea explored in chapter 3 is that some of the advanced characteristics that are frequently associated with HFA depend differentially on certain evolved *right*-hemisphere specializations, whereas the analytical skills associated with AS make use of processing networks that evolved particularly in the *left* hemisphere. As we will see, although HFA is not an officially recognized category, much of what this book reveals about Aspies' likely evolutionary trajectory may well apply to high-functioning autists who have relatively enhanced right-hemisphere visuospatial skills—folks who thought, and think, in pictures, to quote Temple Grandin.[24] Continuing this thread, hitherto underappreciated differences between boys and girls with AS or HFA are detailed and considered from an evolutionary perspective in chapter 4.

I hasten to point out that the origin of the three driving evo-devo trends described in the first chapter had nothing to do with AS or HFA per se, which, instead, emerged much more recently on the evolutionary coattails of these trends. These coattails had genetic underpinnings, of course, and these are discussed in chapter 5, along with relevant clues gleaned from studies of ancient DNA extracted from prehistoric human remains. Chapter 5 also examines the intriguing suggestion that the incidence of AS and HFA may be rising in certain communities like Silicon Valley because of increasing marriages between highly talented technical-minded people, which we respectfully call the "geek hypothesis."[25] The penultimate chapter takes stock of the cross-cultural manifestations of autism at a global level. Most, but not all, of the relevant studies from developing countries are about broader forms of autism rather than AS and/or HFA specifically. These studies prove to be surprisingly helpful when thinking about evolutionary pressures that may affect autism, as well as how such pressures might change and alter its global prevalence in the future. Putting all of this together in chapter 7 leads to the startling conclusion that when it comes to the evolutionary dynamics that influenced the emergence of AS and HFA, and the extent to which autism now exists around the world, we ain't seen nothin' yet.

The evolutionary perspective of this book must not detract from our awareness of the social difficulties that accompany AS and HFA. As is well known, Aspies' intensely focused and narrow interests are associated with physical and social trade-offs. For example, like the stereotypical nerds in popular culture, Aspies tend to be unathletic and socially awkward or clueless. And just like the "geeks" with slide rules in their pockets that my generation grew up with, they are all too often subjected to devastating bullying from peers. I am writing this book with Eve not only because her perspective illustrates and underscores the scientific information but also because she provides an insider's look at the experience of having AS in a society that has not yet come to grips with the condition. In keeping with the neurodiversity movement, we hope that exploring Asperger syndrome and HFA from an evolutionary perspective will contribute to an appreciation of these conditions not only as natural human variations but also as ones that benefit society—and always have.

Before Asperger Syndrome

Three Evo-Devo Trends that Made Us Human

> In the treatment centre for autistic children in Tokyo ... female assistants
> carry small, often highly disturbed children on their backs....The theory is
> that the children receive a certain satisfaction from "skin contact."
>
> —HANS ASPERGER[1]

Although many people associate human evolution with cavemen, like Neandertals, or with archaeological discoveries such as the stunning cave art from Europe, Neandertals and cave art emerged far too recently to shed light on the origin of our prehistoric ancestors, the first hominins. For that, we must go back millions of years earlier in Africa, when the hominin lineage diverged from the predecessors of our closest living cousins, the chimpanzees. To capture glimpses of how and why hominins evolved, paleoanthropologists study their fossilized bones and compare them to those of modern apes and humans. They also observe the changes that occurred in hominin skeletons through time and infer prehistoric behaviors by interpreting the functions of different skeletal anatomies in apes and humans.

Evidence from the growing field of evolutionary developmental biology, known for short as evo-devo, suggests that modifications in the development of fetuses and infants frequently triggered the branching of evolutionary trees. Evo-devo researchers, therefore, compare how infants of different animals develop in order to explore their evolutionary histories and how they emerged as distinct species. When combined with studies of fossils, evo-devo research is extremely powerful.

For example, nearly half a century ago, "man the hunter" was proposed as the main driver of human evolution.[2] Feminist anthropologists responded some years later by nominating "woman the gatherer" as an alternative.[3] Although no one disputes the prehistoric significance of hunting and gathering food, both ideas are now being eclipsed by increasing evidence from evo-devo research, which suggests that ancestral babies,[4] rather than grownups, may have facilitated the most important advances in human evolution. Specifically, evo-devo researchers are shedding startling light on the evolution of prehistoric hominins by comparing development in the fetuses and infants of chimpanzees and humans. This research reveals that three evo-devo trends that began in prehistoric babies became profoundly important for making us human, which suggests that "baby the trendsetter" may, indeed, have been a more significant driver of human evolution than either "man the hunter" or "woman the gatherer."[5]

Before describing these trends, however, we need to clarify what it means to be human. Here, this concept refers to the behavioral and cognitive traits that distinguish people from the other primates, particularly the great apes (orangutans, gorillas, and chimpanzees). Although some scholars believe that tool production, hunting, making war, or social intelligence made us human, these suggestions are unconvincing because some of the other primates also do these things. The traits that truly made us human are more likely to be revealed in how we think. Because people are the only primates that produce and comprehend extremely complex symbolic and grammatical speech, it is highly likely that the emergence of language nudged our ancestors along the road to humanity. Humans are curious about the world around them and continually ask questions (Where did I come from? Why did that apple fall from the tree? Is anybody else out there?), something even the most adept language-trained apes never do. The emergence of intellectual curiosity should thus be included in the traits that made us human. People also take pleasure in figuring out how things work and producing inventions or other creative innovations (symphonies, paintings, mathematical proofs, choreographed dances) that require analytical, systematic, and/or creative thinking that go way beyond the capabilities of other primates. These proclivities are part of our humanity, as is the unique human predisposition for daydreaming, mind wandering, and worrying about the future. In a nutshell, evolved brains and advanced minds are what made, and make, us human.

Bodies evolved too, of course. As we will see, the first changes that distinguished the bodies of hominins from apes facilitated the evolution of walking on two legs (bipedalism), which paved the way for the much later emergence of big brains and advanced cognition (see fig. 1.1). Although Asperger syndrome

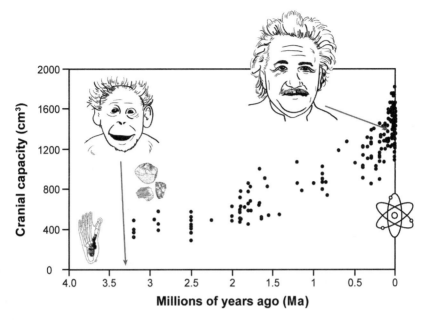

Figure 1.1. The evolution of upright walking, brains, and culture. Bipedalism may have begun to evolve about seven million years ago (Ma), long before about 3.0 Ma, when brain size started to increase. Between the time of the australopithecine infant shown here and Albert Einstein, brains evolved in size and internal connectivity, and material culture expanded from simple stone tools to the products of the atomic age. Brain size is indicated by the cranial capacities of braincases (cm3). Based on a graphic from Nick Matzke of the National Center for Science Education (http://ncse. com/creationism/analysis/transitional-fossils-are-not-rare). Reproduced from Falk, "*Australopithecus* to Albert Einstein."

is a developmental condition that affects modern infants, it is important to emphasize that autism did not exist this early in prehistory.[6] In fact, it would take several million years for the advanced cognitive traits associated not only with AS but also with the special features that made us human to emerge on the coattails of three evo-devo trends that began much, much earlier.[7]

The First Trend: Late Bloomers

It is commonly estimated that the earliest hominins diverged from the ancestors of chimpanzees between five and seven million years ago,[8] when they began walking more often on two rather than four legs, perhaps because it improved their ability to find food.[9] At that time, our predecessors had a feed-as-you-go foraging and scavenging lifestyle during the day but still regularly climbed trees and built sleeping nests in them each night, similar to living chimpanzees.[10] As hominins gradually reduced the time spent in trees in favor of walking on the ground, profound changes occurred throughout their bodies. The anatomy of modern humans retains these changes, which include a head that is perched directly on top of the neck, tops of the arms that hang by the sides of the shoulder, S-shaped curves in the spine, short and broad pelvises, centrally located knees, and arched feet with modified toes. These transformations did not happen overnight, however. Hominin fossils reveal that an efficient humanlike gait may not have been perfected until about three million years ago.[11]

The fossil record reveals further details about the anatomical changes that accompanied upright walking and the order in which they occurred. The feet began evolving first, when they lost the ability to grasp branches and became the weight-bearing organs necessary for upright walking. You can get an idea about how the feet evolved by comparing your hands and feet. Hold your hands out, spread the fingers and thumbs as far as possible, and wiggle them. Now try it with your feet. As you can see, fingers have extensive flexibility for grasping but, thanks to our ancestors becoming bipedal, our toes do not. This is because, compared to their ancestors, hominin's big toes became longer, stronger, and aligned with the other toes, which, subsequently, became shortened and less curved.[12] Rather than grasping, the big toe's new job was to propel walkers forward with each step, after the weight of their body shifted along the sole of their leading foot from the heel to the toe—so-called toeing off. As you can see, none of your toes is likely to be of much help if you want to hang onto something, pick it up, or manipulate it. Other primates have feet that are much more like hands, which they need because they spend a good deal of time (some species, round the clock) moving in trees or on cliffs.[13]

In addition to direct evidence from fossils, studies of motor reflexes in newborn humans suggest that the loss of flexibility in hominin feet was the first domino in a drawn-out series of prehistoric evo-devo changes that, over time, facilitated the evolution of efficient bipedalism. Human babies are born with a grasping reflex in their feet (and hands) that can be elicited by placing a finger underneath their toes.[14] This reflex, which is a rudiment of "responses that were once essential for ape infants in arboreal life,"[15] becomes neurologically inhibited by the time modern babies begin to walk, typically between six and twelve months of age.[16] Natural selection in prehistoric infants, thus, inhibited reflexes that were once adaptive for arboreal locomotion and elaborated others that were preparatory for walking (e.g., babies' stepping reflex).[17] These evolutionary tweaks prolonged the duration of infants' overall motor development.[18] In other words, rather than a neat replacement of one form of locomotion by another, selection for bipedalism was achieved by modifying and adding complexity to existing neuromuscular processes, which increased the time it took to develop the skills needed to become bipedal. This pattern has been inherited by modern humans, which explains why our babies are slower than chimpanzees to achieve milestones like pushing off, sitting, crawling, standing, and walking.[19] The first evo-devo trend was thus for delayed physical development (see fig. 1.2). Put more simply, prehistoric babies became late bloomers.[20]

The loss of grasping feet dramatically modified the trajectory of hominin evolution. Soon after birth, chimpanzees develop an ability to cling with all fours to their mothers' bellies, where they are content to nurse and ride for long periods without assistance as their mothers move about on the ground or in trees. When they are a bit older, chimps shift to riding on their mother's back. After hominin feet became adapted for walking at the expense of grasping, prehistoric babies no longer developed the ability to cling unassisted to their mothers' bellies and backs, which meant that mothers had to help them stay attached. The loss of independent riding was a remarkable evolutionary reversal because all monkey and ape infants develop this skill.[21] Although it may seem odd to think of this reversal as part of an evolutionarily advanced package, it was. The shift from unassisted to assisted riding coupled with infants' late-blooming locomotor milestones increased the amount of time youngsters were dependent on caregivers and, consequently, their opportunities for learning. Infants' increased helplessness, thus, may have contributed to the strengthening of social bonds, not just between infants and mothers, but also within groups.[22]

Because bipedalism deprived prehistoric infants of grasping feet, mothers had little choice but to carry them in their arms or on their hips as they made their daily treks in search of food, perhaps with help from other caregivers. Carrying infants in this way is extremely hard work,[23] which is why anthropologist Adrienne Zihlman and others suggest, quite reasonably, that

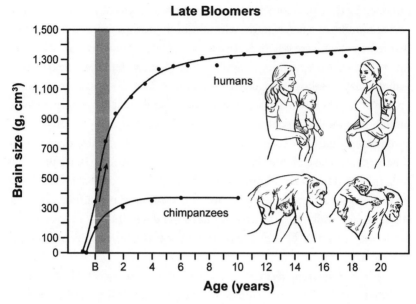

Figure 1.2. The first evo-devo trend: late bloomers. Average brain-growth curves (brain size in grams [g], roughly equivalent to cm3) are plotted against developmental age for humans and chimpanzees from birth (B). Because it takes longer for the brains of humans to mature, human babies are delayed in physical development and thus slower than apes to achieve milestones such as crawling, standing, and walking. Consequently, human infants never develop the apelike (and monkey-like) ability to cling unaided to their mothers. As shown, the responsibility for attaching infants to caregivers shifted prehistorically from infants to mothers. Reproduced from Falk, "Baby the Trendsetter."

baby slings were among the earliest cultural inventions.[24] Even so, the fact that bipedalism originated long before the appearance of the first stone tools (5–7 million years ago versus around 3.5 million years ago) suggests that it may have taken quite some time for hominins to invent baby slings. (Interestingly, once they were invented, "man the hunter" may have played a role in propagating the use of baby slings by providing leather for them, although plant material was probably also used. In any event, leather may have been a highly desirable commodity. Did males give it to females during prehistoric courtship? Were leather slings the first ever baby gifts?)

Today, "baby wearing" with slings is universally practiced, although more so in some cultures than others. Nonetheless, carrying infants for extended periods of time gets tiring, even with the use of baby slings. It is common, therefore, for mothers to place babies in hammocks, bassinets, cribs, or playpens, or simply nearby on the ground—preferably on a blanket of some sort. Because other primates do not do this with their nursing infants, the practice of putting babies down must have been invented by our hominin ancestors. Although we take this custom for granted, its invention triggered a second evo-devo trend, in which detached infants developed ways to reestablish contact with their mothers.

The Second Trend: Seeking Contact Comfort

Primate infants, including human ones, hate to be separated from their mothers, even by very short distances. The deep evolutionary roots of youngsters' need to be attached to their mothers was vividly demonstrated in a series of disturbing experiments that were conducted by psychologist Harry Harlow in the 1950s and 1960s. Harlow and his colleagues removed newborn macaque monkeys from their mothers and raised them in isolation chambers in which their only company was a wire-framed surrogate mother that contained a feeding bottle and another cloth-covered surrogate mother that did not. Although the wire mother with the bottle provided all their nourishment, the monkeys spent virtually all their nonfeeding time clinging to the cloth mother. Needless to say, these pathetic infants grew into extremely disturbed adults, which led Harlow to conclude that "contact comfort" was at least as crucial as mothers' milk for the development of healthy infants.[25]

Fortunately, wild monkey and ape nurslings are rarely separated from their mothers' bodies, which is a good thing because remaining attached is of vital importance, as illustrated by what happened to the offspring of a wild chimpanzee mother named Madam Bee:

> Madam Bee had raised two infants successfully when one of her arms was paralyzed during a presumed polio-epidemic. . . . The two infants that were born afterwards died within a few months. I had the occasion to make observations on the first of these two infants: Bee-hind. Her body was full of wounds and scratches, so she must have fallen repeatedly. Whenever her mother moved about without supporting her, she whimpered and screamed continuously.[26]

Poor little Bee-hind's experience suggests that prehistoric hominin infants would have suffered high infant mortality because their evolved feet made unassisted riding on their mothers impossible. Further, hominin babies who escaped Bee-hind's fate would have had diligent mothers who developed methods for coping with their, literally, insecure infants. An obvious way mothers could have prevented or at least lessened the number of falling episodes was to put babies down periodically in a nearby location, preferably within arm's reach. As we have seen, this practice would have had the added benefit of providing mothers some respite from actively carrying their helpless, growing infants.

Before hominins split from chimpanzees, young infants would have remained, for the most part, silently and contentedly attached to their mothers' bodies, similar to living chimpanzees. Once bipedalism was underway, however, babies who fell from their mothers probably expressed distress with vocalizations similar to Bee-hind's whimpers and screams, and those who were put down may have cried in the breathy, tearless manner of chimpanzees.[27] As prehistory continued to unfold, infants and toddlers evolved more elaborate ways for protesting separation from, and trying to reestablish contact with, their mothers, including the use of persistent (some would say incessant) crying. According to British psychoanalyst John Bowlby, the originator of attachment theory, modern toddlers attempt to prevent separation from their mothers not only by crying but also by kicking, using beseeching gestures, or hanging onto a leg.[28] Bowlby was one of the first to theorize that the evolution of such behaviors and other infant characteristics (such as cuteness and appealing facial expressions) stimulated attention from caregivers and thus increased babies' chances of survival. This idea seems especially apt when one considers that crying in human infants, which has unique acoustic features, tends to be sustained, and includes an ability to shed emotional tears, an ability other primates do not have.[29] As anthropologist Meredith Small observes, crying and the parental sensitivity to it

> evolved to serve the infant's purposes: to assure protection, adequate feeding, and nurturing for an organism that cannot care for itself. By definition, crying is designed to elicit a response, to activate emotions, to play on the empathy of another. . . . The caretaker has also evolved the sensory mechanism to recognize that infant cries are a signal of unhappiness, and thus be motivated to do something about it.[30]

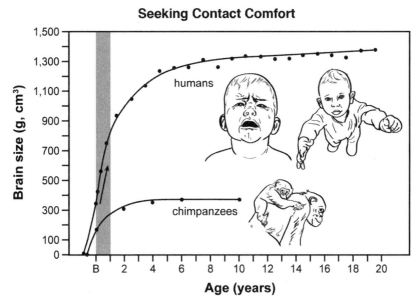

Figure 1.3. The second evo-devo trend: seeking contact comfort. Average brain-growth curves (brain size in grams [g] and cm3) are plotted against developmental age for humans and chimpanzees from birth (B). Prehistoric infants who failed to develop the ability to cling independently to mothers because of the evolutionary loss of grasping ability developed vocal and visual signals (e.g., shedding emotional tears) that prompted caregivers to pick them up or provide other forms of comfort (e.g., soothing vocalizations, rhythmic bouncing). These signals paved the way for reciprocal communication between infants and mothers and the eventual emergence of baby talk (motherese), which seeded the first language. Reproduced from Falk, "Baby the Trendsetter."

In keeping with attachment theory, numerous other studies show that although infants cry for various reasons, a main reason babies from industrialized societies do so is to reestablish contact with caregivers (see fig. 1.3).[31]

The significance of the trend for infants seeking contact comfort from separated caregivers—by fussing, gesturing, persistent crying, shedding emotional tears, and unhappy vocalizations—cannot be overstated. For one thing, I believe it stimulated intense give-and-take gestural and vocal communications between infants and mothers that sparked the evolution of facial gestures (for example, prolonged eye contact, and reciprocal smiling between mothers and infants) and other forms of body language.[32]

In response to their babies' quest for contact comfort, ancestral mothers invented behaviors that caregivers still use to soothe and hush babies, including gestures that simulate comforting sensations experienced when infants ride on foraging mothers (rocking, bouncing, jiggling) and other gestures that reestablish body contact, if only temporarily (snuggling, hugging, picking infants up).[33] The invention of other stimuli that simulate contact with mothers such as blankies, swaddles, and pacifiers would have occurred more recently, after the invention of textiles. Prehistoric mothers would also have invented certain types of calming vocalizations that do not exist in other primates but are widely used by modern parents (hushing, cooing, and murmuring).

As evolution progressed, vocalizations between mothers and infants continued to evolve. For their part, infants' cries evolved melodic components that became more complex over the first few months of life, a pattern that our infants have inherited. Significantly, the sequential development of melodic cries plays an important part for the later appearance of babbling in modern babies, which, in turn, paves the way for learning to speak.[34] As for prehistoric mothers, their soothing coos and murmurs eventually evolved into more melodic intonations, including the first lullabies; these were hummed rather than sung with words, which had not yet been invented. As detailed in my "putting the baby down" theory, which I developed some years ago,[35] the earliest form of baby talk (known more formally as "motherese" or "parentese")[36] blossomed after the first words were invented and subsequently evolved into the more complex forms universally used by today's parents.[37]

Ample evidence from comparative psycholinguistics, developmental psychology, and cultural anthropology shows that, in addition to contributing to the emotional and social well-being of infants, motherese helps babies the world over learn their native languages. This is why the putting the baby down theory suggests that the invention of motherese by prehistoric mothers was the spark that ignited the eventual emergence of the first symbolic language (called protolanguage).[38] As noted, the emergence of language was probably hugely important for our ancestors' journey to becoming human. Some might view the notion that mothers and babies were primarily responsible for inventing language with a jaundiced eye, because such an idea may seem biased against adult males. We need to keep in mind, however, that hominin evolution depended heavily on the individuals who passed their genes to future generations (babies) and, therefore, on those who were most directly responsible for their survival (prehistoric mothers). It is different today, of course, because, in contrast to apes and presumably our early ancestors, people know where babies come from (that is, that they have two parents), and fathers participate in parenting. That said, the invention of language alone would not have been enough to make us human. As noted, what ultimately did was our evolved brains and advanced minds. This brings us to the third evo-devo trend, which entailed an unprecedented growth spurt in the brains of prehistoric fetuses and infants.

The Third Trend: Early Brain Spurt

The fossil record shows that between three and four million years ago, hominin brain size began to increase and that it continued doing so until it leveled off at its modern size, which is about three to four times as large as the brains of our earliest adult relatives and living apes. In the course of this evolutionary increase in size, the brain's internal organization became more complex,[39] and the archaeological record of cultural and technological changes progressed from simple stone tools, at around 3.4 million years ago,[40] to radioactive sediments from nuclear test blasts that introduced the atomic age (fig. 1.1).[41] These dramatic changes were not characterized by gradual growth of the brain over the lifetimes of early hominins but rather by an unprecedented rapid acceleration in brain growth in fetuses and newborns, a phenomenon I call the "brain spurt." Modern babies inherited this spurt and, as figure 1.4 shows, their brains continue to grow rapidly after the first year, though at a decelerating rate until brain size practically levels off in the teens.[42] Brain size increases a little over 100 percent in the first year, but only about 15 percent in the second.[43]

When one examines brain-growth curves in living primates together with those inferred from the fossilized braincases of prehistoric hominins, it becomes clear that the brain spurt, which is strongly manifested in modern human infants compared to apes (arrow in fig. 1.4),[44] was probably the mechanism that tugged brain sizes upward during the course of hominin evolution. Consistent with this, the brain-growth trajectories of living monkeys, apes, and humans are similarly shaped, but nested one above another

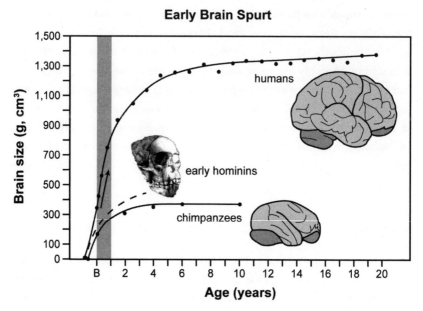

Figure I.4. The third evo-devo trend: early brain spurt. Average brain-growth curves are plotted against developmental age for humans and chimpanzees from birth (B). In humans, brain growth is accelerated before and after birth and is especially noticeable (arrow) during the first year of life (shaded). This trend accounts for the fact that humans grow brains that are more than three times the size of adult ape brains. The dashed line illustrates the evolutionary emergence of an accelerated brain spurt in *Australopithecus*, represented here by the approximately 2.5-million-year-old Taung child. The brain sizes for chimpanzees and Taung are estimated from braincase volumes (cm3); those for humans are brain mass in grams (g). Data for humans and chimpanzees from Passingham, "Changes in the Size." Reproduced from Falk, "Baby the Trendsetter."

as the curves progress from monkeys to apes to humans—illustrated here by the plots for chimpanzees and humans.[45] Inferred curves based on braincases from fossil hominins such as Taung (shown on the graph) are consistent with the suggestion that brain spurts of increasingly big-brained hominin species provided the impetus that edged adult brain sizes higher and closer to the human curve as hominins evolved through time from australopithecines to *Homo habilis*, *Homo erectus* and, finally, *Homo sapiens*.

The brain spurts of modern infants are driven mostly by increases in the number of connections between various parts of nerve cells that are located in the gray matter, including the junctions (synapses) between the cells that transmit neurochemical impulses.[46] Synapses do not just increase during the brain spurt; many are also eliminated as the nervous system becomes tuned to the individual's particular environment.[47] To a much lesser extent, brain spurts also include a rise in white matter, which consists of nerve fibers that are insulated with a pale substance called myelin that increases their conduction speed.[48] It is a good bet that similar increases in gray and white matter occurred during brain spurts of our prehistoric relatives.

Whatever caused the brain spurt to emerge and continue evolving in our ancestors must have had a significant evolutionary payoff because its retention in humans comes at great cost for women during childbirth. This is because babies' heads (brains) have evolved to be relatively large compared to their mothers' bony birth canals. This so-called obstetrical dilemma happens because modern women have pelvises (and therefore birth canals) that were shaped millions of years ago to accommodate the muscle arrangements required for bipedalism—a design that obviously did not leave enough wiggle room for modern-sized babies to be easily born, unlike their relatively comfortably born ape cousins.[49]

In order to explore the evolutionary causes of the brain spurt, we can ask what happens to human babies during this stage of development that does not happen to chimpanzees. Experiments have shown that human fetuses spend their last trimesters eavesdropping on their mothers' speech, which they have a strong preference for listening to after they are born.[50] In other words, babies perceive critical aspects of their native languages "beginning *in utero*, which continues through the first year of life before they can speak."[51] During babies' first year, they learn how to recognize the smallest speech sounds (phonemes), each of which is coded in multiple parts of their temporal lobes near other regions that perceive sound more generally.[52]

Developmental psychologist Patricia Kuhl and her colleagues have shown that by six months of age, infants are able to distinguish all of the approximately eight hundred speech sounds in the world's collective languages, but by the end of their first year, their ears and brains have become tuned selectively to just the forty or so speech sounds in their native languages.[53] Remarkably, once infants reach this point, they use unconscious statistical computations to discover words in running streams of speech. As babies continue to mature, their brains form widespread neurological networks that help them understand the meanings of words, as well as how to assemble them with correct components (such as endings that make them plural or past tense) and into correct phrases and sentences (called syntax).[54] Infants develop these aspects of grammar by listening to speech—especially motherese, which exaggerates salient features of language—while watching what is happening, such as when Baby sees Mother patting Fido on the head while crooning "nice doggie."

Grammatical language is computational, combinatorial, hierarchical (i.e., phonemes are embedded within words, which are embedded within phrases, and so on), and sequentially processed. Above all, it is a symbolic system that allows humans to experience and communicate an endless variety of ideas. However, it would be a mistake to think of the increasingly wide neural networks that infants begin to develop prenatally as webs of connected neurons that are dedicated exclusively to language.[55] These networks have vast connections, particularly within the left hemisphere,[56] which make them available for processing other kinds of systematic thinking as babies grow up, such as that related to music, art, mathematics, and science.[57] For example, statistical processes similar to those used to acquire language facilitate infants' learning of visual patterns.[58] Similarly, although written words look very different in Spanish, English, Hebrew, and Chinese, the "reading network has evolved to be universally constrained by the organization of the brain network underlying speech."[59]

As noted in the introduction, the right side of the brain develops specializations that differ from those of the analytical left hemisphere. Neuropsychologist Allan Schore of the University of California at Los Angeles believes that infants' right hemispheres mature in response to interactions with caregivers. Thus an infant's "interactively regulated right brain . . . attachment experiences become more holistically integrated, allowing for the emergence of a coherent implicit (unconscious) emotional and corporeal sense of self."[60] This fits well with the emphasis on reciprocal mother-infant interactions in the putting the baby down theory, and suggests that development of the strong right/left differences that characterize human brains was and is part and parcel of the brain spurt.[61] We will return to this topic, including the possible influence of right-hemisphere evolution on the emergence of visual thinkers with high-functioning autism, in chapter 3.

Whence Asperger Syndrome?

As any parent with more than one child knows, individual babies express the three evo-devo trends to different degrees. Some achieve milestones like sitting up or walking later than others, others cry more than average and may be needier when it comes to contact comfort, and some babies are born with relatively larger heads (brains). In other words, although these three trends clearly set humans apart from the other primates, there is a certain amount of natural variation in their individual development. The thing to keep in mind is that the three trends were evo-devo dominoes that, over vast amounts of time, resulted in the emergence of advanced human cognitive abilities. Hans Asperger's famous quip that a dash of autism is essential for success in science or art (quoted in the introduction) raises the enticing question of whether these trends are expressed any differently in Aspies than in typically developed (TD) children.

As detailed below, Aspie infants appear to be on the high end of the range of variation for the first and third trends compared to neurotypical babies and express the second trend in fascinating ways that accommodate their highly unusual sensory systems (a subject discussed in the next chapter).[62] Regarding the first trend, people with AS tend to be later bloomers than TD individuals when it comes to developmental milestones like sitting, creeping, walking, and so on.[63] For example, one study found that the average age for walking in eighty-five males with AS was 13.8 months, compared to 12 months in the general population.[64] Aspies' delayed motor development may be part of the reason they tend to be physically clumsy and have poorer coordination of their two hands, even compared to individuals with HFA.[65] Remarkably, some Aspies seem to shed their autistic symptoms by the time they become adults, although this is sometimes incorrectly attributed to a "cure"[66] rather than recognized as a possible manifestation of the evo-devo trend for prolonged development.

The fact that individuals with AS tend to be more delayed in their motor development compared to neurotypical children fits with the observations of evolutionary biologist Bernard Crespi, of Simon Fraser University, who is one of the few researchers who views autism within an evo-devo framework. In addition to observing that "the differences between autistic and typically developing individuals mirror the differences between younger and older typically developing individuals," Crespi showed that several core autistic features are associated with altered timing and delays in the development of individuals.[67] More to the point, he recognized that the "evolutionary extension of [delayed] child development along the human lineage has potentiated and structured genetic risk for autism and the expression of autistic perception, cognition and behavior."[68] The views developed in this book about the relationship between the first evo-devo trend and the eventual emergence of AS are consistent with Crespi's findings.

Regarding the second trend (seeking contact comfort), many children with AS and HFA are calmed by the same kinds of sensory stimuli that our ancestors administered to comfort helpless infants, but because of hypersensitivities detailed in the next chapter, they find it difficult to tolerate parents or others delivering these stimuli. As autistic scientist and author Temple Grandin put it, "I craved tender touching. I ached to be loved—hugged. At the same time I withdrew from over-touch. . . . For me and many autistic children . . . our bodies cry out for human contact but when contact is made, we withdraw in pain and confusion."[69] Many autistic children, including those with AS, thus prefer to administer their own hugs (Grandin's famous squeeze machine, for example) and to swaddle themselves in tight clothes or commercially available weighted blankets that are designed to provide soothing pressure for autists.

Similarly, Aspies may experience comfort from being in tight enclosures or by stroking soft materials such as animal fur. They also frequently seek relaxing rhythmic stimulation by bouncing, flapping limbs, spinning, or rocking themselves. (Significantly, monkey infants raised in social isolation developed behaviors similar to these).[70] Unfortunately, these unusual mannerisms are all too often stigmatized as undesirable "stimming" rather than recognized as remarkable adaptations that serve an evolutionarily engrained need for contact comfort.

Until recently, it was thought that autistic infants did not respond favorably to baby talk,[71] but recent analyses of home videos are changing that. French psychiatrist David Cohen and his colleagues found positive responses to parentese in pre-autistic babies and that mothers and fathers of babies who were later diagnosed with autism may have unconsciously sensed that parentese increased desirable responses in their infants. To my delight, the authors noted that their findings were consistent with the putting the baby down theory. Although the home video studies were not focused on children who later developed AS,[72] it is heartening that clinicians are beginning to recognize the potential therapeutic value that exposure to baby talk may have for treating autism.[73]

With respect to the third trend, it is well known that early brain growth is accelerated in some of the babies who will later be diagnosed with autism, resulting in relatively enlarged brains during childhood (but not later) compared to TD infants.[74] Traditionally, researchers have measured head circumference as a proxy for infants' brain size, although it is now state-of-the-art to determine actual brain volumes from magnetic resonance imaging (MRI). For example, one recent study used MRI to obtain brain volumes from one hundred males with Autistic Spectrum Disorder aged three to thirty-five years, with many subjects imaged several times as they grew up, and compared these volumes to those from fifty-six TD males.[75] This study confirmed that in autism writ large, the trend for average brain growth in youngsters accelerates above the curve for TD individuals, then subsides until it crosses beneath the TD curve at between ten and fifteen years of age, after which brain size in autists continues to decrease for some time before leveling off (fig. 1.5).[76] Brain size in TD individuals shows a similar pattern, but the decrease in brain size begins after age fifteen or sixteen.[77] The fact that some ASD infants have significantly enlarged brains suggests that brain growth in autistic infants overlaps the high end of the range of variation for TD infants.

A number of researchers have studied the microscopic structure of post-mortem infant brains in order to learn possible reasons for the relatively enlarged brain spurts of some autistic babies compared to neurotypical infants. Recently, brain overgrowth in autism was found to involve hyper-expansion of just the cortical surface during the second half of the first year, followed in the second year by overgrowth of the entire volume of the brain.[78] Other studies of autists' brains describe an overabundance of, and less spacing between, neurons in certain regions, particularly in parts of the frontal and temporal lobes.[79] In fact, one study suggested that, given the excessive number of cortical neurons in autistic children, their brains should have grown to be even larger.[80] The fact that cortical neurons are generated before rather than after birth suggests that the excess number of neurons and relatively large brain spurts seen in some autistic individuals is likely due to alterations in prenatal developmental factors.[81] Interestingly, the brain spurt of TD humans begins around the twenty-second week of gestation,[82] which hints that at least some of the neurological aspects of autism may, indeed, begin surprisingly early during prenatal development.[83]

But where are the Aspie infants in all of this? Thanks, in part, to the removal of Asperger's Disorder from *DSM-5*, they are lumped with other autistic individuals on the ASD brain-growth curve in figure 1.5.[84] The best current estimate is that over one-tenth of individuals diagnosed with subtypes of ASD have AS.[85] So far, there are few if any studies that have examined either cortical microstructure or brain growth in infants with AS. Similarly, very few studies offer clues about how growth of the whole brain in Aspies compares with brain growth in other autistic individuals. One that does, however, found that a significantly enlarged head circumference "is rather more typical of the highest functioning variants of autistic spectrum disorders, currently referred to as Asperger syndrome."[86] This finding fits well with the observation that "autistic individuals with accelerated head growth in early childhood displayed higher levels of adaptive functioning and less social impairment"[87] compared to autists without such an acceleration. Despite their initial differences, the brain-growth curves of Aspies and neurotypical individuals eventually seem to even out since "previous studies found no differences in the total brain volume between adults with Asperger syndrome and controls."[88]

The postnatal acceleration in brain growth of human babies has long been recognized as an advanced evolutionary trend.[89] This suggests that the conventional wisdom that the early brain spurt in Aspies (or people with HFA, for that matter) is pathological (i.e., a sign of disease) needs rethinking.[90] In fact, since the brain spurt in neurotypical infants is associated with their development of distributed neural networks, especially in the left hemisphere, which process linguistic and, eventually, other kinds of abstract computational thinking, one suspects that the pronounced early brain spurt of Aspies may play an important role in kindling their particular intellectual strengths.[91]

In this context, it is interesting to view AS from the perspective of the growing neurodiversity movement, which asserts that atypical neurological development is a natural component of human variation.[92] Some believe the concept of neurodiversity is problematic when applied to low-functioning autism, but ethicists Pier Jaarsma and Stellan Welin make a compelling argument that the neurodiversity movement got it right when it comes to high-functioning autism. In their opinion, rather than being a disorder, most

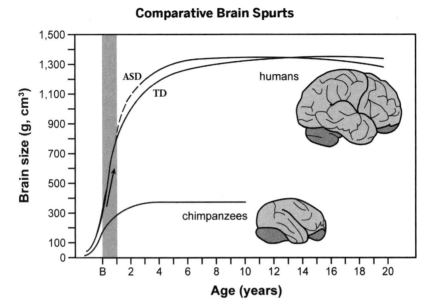

Figure 1.5. Comparative brain spurts. Average brain-growth curves (brain size in grams [g] and cm3) are plotted against developmental age for autistic (ASD) and typically developed (TD) humans and chimpanzees from birth (B). The early brain spurt is more accelerated, on average, in ASD than TD individuals. Prebirth curves for TD individuals and chimps after Sakai et al., "Fetal Brain Development"; neo-natal parts of curves for TD humans and chimpanzees from Passingham, "Changes in the Size." Curve for ASD males older than three years based on Lange et al., "Longitudinal Volumetric Brain Changes." The dashed part of the ASD curve is inferred from the literature. Some researchers have concluded that the brain-size trajectories for Aspies and TD individuals reach a similar level later in life. Prepared by Tony Archer.

of the problems of AS may be due to social conditions associated with "the 'autism-phobic' character of present society."[93] (We will return to this in chapter 6.)

The evolutionary perspective of this book is consistent with such a view because Aspies are, if anything, ahead of the curve when it comes to the advanced trends that were sparked by prehistoric babies—trends that made us the uniquely talkative and mentally computational species we are today. Shouldn't that be enough reason to recognize AS as a condition that is culturally constrained by social vulnerabilities rather than as a mental disorder?

After all, our species is currently undergoing an almost unimaginable transformation in information technology, and it is a good bet that Aspies with high levels of expertise in narrowly focused subjects (some would say obsessions) will be overrepresented among the innovators who move us forward, not only scientifically but also in other creative pursuits that make use of distributed computational neural networks.

Eve is among those Aspies who have such obsessions. In her case, a precocious reading ability triggered the development of her interests. Although reading emerged in humans only about fifty-five hundred years ago, it piggybacked on the neurological networks that evolved much earlier in association with the prehistoric evo-devo trend for an early brain spurt (more on this in chapter 7.) As is typical of Aspies, Eve's interests are frequently manifested in a repetitive motor mannerism (stimming). However, rather than using stims such as hand flapping or spinning in circles, Eve talks to herself out loud when she is alone. In keeping with one of the themes in this chapter, Eve's following narrative makes it clear that her form of stimming not only satisfies a deeply engrained need for soothing contact comfort in the form of self-administered baby talk (related to the second evo-devo trend) but also reflects her particular intellectual focus.

EVE: MY OBSESSIONS

One of my earliest memories is making trails out of the books on Dad's shelves. Another early memory is that my favorite CD-ROM was called Reader Rabbit and, according to my mother, it was that CD-ROM that helped me to teach myself how to read at age four. While a lot of the kids in my class were still reading some books with a few chapters and many pictures I was reading books with multiple chapters and few illustrations. If I had to pick an event that really started my interest in spelling, vocabulary, and the subject of English in general it would have to be the fourth-grade spelling bee. That was when my class, which had been doing in-class spelling contests for a while, went up in front of the entire school. There we were, standing on the risers in the auditorium at school. The principal was calling out the words and we were taking turns going to the microphone to spell the word. It finally came down to just me and one other girl there on the risers. The problem was that we had run out of words from the list that we were given to study the night before the contest. The principal sent one of the teachers to run for a dictionary so that they could find new words. The first person to get the word wrong was out of the contest. I won on the word "consequence." When asked how I knew the spelling of such a big word, I could truthfully say that I just saw it spelled often enough on a poster in my favorite room in the school—the library.

Today, I carry four books in my arms wherever I go in public. There are several reasons I do this. The first being that books are basically my barrier against the world. Even in my classes at Bath Spa University I carried four books for reading for fun in addition to the books required for class. Books that were required for class I carried in my backpack. The second reason I carry four books is that I like to read and I am a fast reader. The upshot of this is that I finish books long before other people do and I would rather have another book to read than have to sit around being bored. I guess you could say that reading is my hobby and four books just seems like a nice safe number to carry around. I know there is no way that I'll be able to finish all four books in one day, but any less than four and I tend to feel as though I'll run out of books to read and that makes me nervous about having to inter-act with the world that is so full of different noises that it is hard to make out voices. In answer to the question of whether I carry four books because I am inherently superstitious, I would say that I really am not apart from a slight belief in fairies that stems from my stepfather Michael's insistence on telling me I was a fairy when I was a child. (This tale of the changeling is an old belief that fairies used to steal children from houses with open windows and leave a fairy baby in its place.)

Most Aspies have some sort of enhanced interest in one or two particu-lar subjects that usually starts to border on avid obsession. In my case I can safely say that anime, which are Japanese animated cartoons, and manga, which are Japanese comics read right to left, are my main obsessions. Any time I don't spend either watching anime or reading manga is usually spent reading fantasy or paranormal romance, which it could be said is a subgenre of fantasy. The thing about anime that really appeals to me is the eye-catching color of it all and seamless movement between scenes. There are usually no slipups or hesitations in the computer animated world unless it is part of the story, and even then such slipups and hesitations look natural or at least uncontrived. In most cases I prefer the computer animated anime to the hand-drawn, though I must say that I couldn't do that sort of thing myself because I just can't seem to wrap my head around the ins and outs of a complicated computer program without a lot of help from an expert in that sort of work. I am also caught by the plots of individual series that I watch. In the main, I mostly watch classic card game or tournament or magical girl anime. Though I do on occasion watch mecha [mechanical, robotic] anime. Anime makes me feel as if I have been transported to another world that doesn't overwhelm me. The music usually sets the general atmosphere and helps to draw me into the anime just as much as the lines that the characters are saying do. I have learned many Japanese words from watching English subbed anime.

My obsession over these Japanese art forms goes to the point that I use the plots of the ones I like the most to enhance stories I tell to myself out loud, usually in a volume that I am sure other people aren't going to hear. After all I don't want to be known as that crazy girl who talks to herself. When I'm alone, I up the volume of what I say. Sometimes when I think I'm talking to myself at a volume that only I can hear, I suddenly have someone else, usually a family member, answering whatever question or phrase I have just said or asking me whether I am talking to them. To this I usually answer that I am talking to myself, mostly because it is the truth, but also because the stories I tell during those times are just for my consumption alone and aren't meant for other people to hear or even know about. It does on occasion happen that I tell stories to myself purely in my head without, as far as I can tell, any sound whatsoever coming out of my mouth. At those times, if I look in the mirror I do tend to see that my lips, however, are still moving. That of course tends to tip off people around me that I am indeed talking to myself or at least reminding myself of something. As for when I am talking in my sleep, I really don't know what I'm talking about then. I do know that I can't get to sleep without telling myself at least one story. My mind is too active otherwise, and the story helps me to wind down.

If this doesn't sound too narcissistic, I would also like to say that the sound of my own voice is soothing to me and helps me to figure out problems or situations. I talk to myself so that my logic can keep up with my senses or something like that, or maybe it is just that the talking to myself helps me to make sense of the stimuli in the world around me. Also, sometimes when I talk to myself I use childish or babyish words, often for animals and such like "kitty" or "duckies," usually in an excited tone of voice. This is mostly because animals are so much more fun than humans for me, plus they don't demand I talk to them or pay attention to them. I also get told off for doing that in front of my father, who thinks I shouldn't say words that are childish, though his phrase is usually "don't use baby talk."

Sensory Experiences

A Missing Link for Understanding Asperger Syndrome

Over-sensitivity and blatant insensitivity clash with each other.... In the sense
of taste we find almost invariably very pronounced likes and dislikes.... It
is no different with the sense of touch. Many children have an abnormally
strong dislike of particular tactile sensations, for example, velvet, silk, cotton
wool or chalk. They cannot tolerate the roughness of new shirts, or of
mended socks.... There is hypersensitivity too against noise.

—HANS ASPERGER[1]

When Asperger's Disorder was still listed as a diagnosis by the
American Psychiatric Association, its diagnostic features were
defined in strictly behavioral terms that included impaired
social interactions and restricted repetitive and stereotyped movements.[2]
These behavioral criteria, however, failed to recognize the core importance
of sensory experiences for Asperger syndrome and other forms of autism.[3]
Aspies, for example, tend to be overly sensitive to some kinds of stimuli, and
their reactions can go much further than simply disliking the feel of sand
or itchy sweaters, or fearing the sound of roaring dinosaur exhibits or loud
thunderstorms, which are among the experiences that Eve will discuss below.
Senses, after all, set the stage for conscious awareness: "The brain, the organ
that is responsible for your conscious experience, is an eternal prisoner in the
solitary confinement of the skull . . . and must rely on information smuggled
into it from the senses . . . the world is what your brain tells you it is, and the
limitations of your senses set the boundaries of your conscious experience."[4]

Despite the autism community's focus on behaviors, at least one investigation has compared how children with AS and other types of autism control their exposure to sensory stimuli such as noise, bright lights, touch (for
example from wearing shoes or socks), and sensations in the mouth (such as

the textures of certain foods).[5] The study also examined emotional responses to sensations, fears that interfered with daily life, and difficulties tolerating change. Intriguingly, children with AS reacted significantly more emotionally than other autistic youngsters and had a greater tendency to become distracted because of external stimuli, especially sounds, which led the authors to conclude that "sensory overload [is] more likely to occur with persons who have AS," and that this may be "related to the tantrums, rage, and meltdowns often exhibited by children and youth with AS."[6]

These remarkable findings cry out for an explanation. Why would otherwise high-functioning Aspie children be significantly more sensitive than other autistic youngsters in these particular ways? Perhaps AS is more compatible than other forms of autism with the so-called "intense world syndrome," in which "the autistic person is an individual with remarkable and far above average capabilities due to greatly enhanced perception, attention and memory" and has a "brain [that] needs to be calmed down."[7] Additionally, compared to other children with autism, those with AS have a greater awareness that they are different from others, which probably adds to their emotional frustration.[8]

The relatively high IQs of some people with AS may also lead to a distracting, almost hypnotic, fascination with natural visual phenomena, like the structures of individual grains of sand, or patterns of light, a tendency that has been associated with creativity and discovery in neurotypicals. Scientists are likely to call this mental state "flow," whereas athletes describe it as "being in the zone."[9] Whichever term you prefer, Aspies frequently experience intense focus to the point of distraction. According to Rudy Simone (author of *Aspergirls*), for example, "When we are in *the zone*, we . . . have a hard time with taking breaks, going to the toilet, eating, drinking, grooming, getting fresh air, or exercise."[10] Eve says the same thing.

Unusual sensory processing and emotional regulation are as core for adult Aspies as they are for youngsters with AS. This was shown by researchers who went straight to the source by analyzing the autobiographies of sixteen autistic writers and interviewing five additional French speakers who were autistic, all between the ages of twenty-two and sixty-seven.[11] All the individuals in the study pointed out that "unusual perceptions and information processing, as well as impairments in emotional regulation, were the core symptoms of autism."[12] They also referred to difficulties processing information from more than one sense at a time and, similar to autistic children, many of the adults reported that they disliked being touched, wearing certain clothes, grooming their hair, and brushing their teeth. Some avoided specific foods because of their texture, color, or taste. They described their unusual sensory and perceptual experiences as not only causing great distress, anxiety, and fear, but also as sources of pleasurable fascination, for example with certain lights, movements, patterns, and smells (or, as Eve mentions below, with curly hair and soft fur).

When sensations get to be unbearable, Aspies and other autists with high-functioning autism may modulate their responses by intensely focusing on something that is distracting (getting in the "zone") or by engaging in comforting behavior, as Temple Grandin describes: "My hearing is like having a sound amplifier set on maximum loudness. My ears are like a microphone that picks up and amplifies sound. I have two choices: 1) turn my ears on and get deluged with sound or 2) shut my ears off. . . . I discovered that I could shut out painful sounds by engaging in rhythmic stereotypical autistic behavior."[13]

Rudy Simone, among others, calls self-stimulating behaviors like the ones Grandin mentions "stimming," which she describes as something autistic people do "to *soothe* ourselves when we are upset, anxious, overloaded, or in pain." Simone identifies these behaviors as typically including "rocking, swaying, twirling, spinning (yourself or objects), humming, flapping, tapping, clapping, finger flicking, and so on."[14] She also lists pleasant stims that "happen because of an overflow of positive emotion: dance for joy, laugh out loud, flap my hands or arms, say 'yay' and 'yippee' like a little kid, jump up and down, ball up my fists and shake them, clap, can't stop laughing, pace back and forth at full speed, skip, sing, speak in a high squeaky voice."[15]

Perhaps the best known stim, thanks to Temple Grandin, is the self-administered pressure from tight wrappings, heavy coverings, or a mechanical device such as her famous "squeeze machine." As Grandin recalls, "When I was six, I would wrap myself up in blankets and get under sofa cushions, because the pressure was relaxing."[16] Grandin also realized at a young age that occupying tight and warm enclosures was comforting:

Another idea I had in grade school was to build a small enclosure about three feet wide and three feet tall—just big enough so I could get into

it and close the door. This miniature enclosure would be heated . . .
Warmth and pressure tend to lessen arousal. Perhaps if I had had a magical comfort machine, I could have used its warmth and pressure instead
of throwing a temper tantrum.[17]

Like Grandin, many Aspies associate small enclosures and tight wrappings with feelings of safety or security. Recalling a time when he was five or
six and had crawled under a desk, Aspie author Luke Jackson wrote, "I like
being underneath things. It gives me a feeling of warmth as if I am safe.
The confusing world seems remote and surreal when I have something over-
head or am enclosed in something."[18] When he was young, Jackson would
wear a close-fitted hood that covered most of his head: "I used to wear it
twenty-four hours a day. . . . It was more than just a comfort. . . . It was as if
I was somehow watching this confusing world from behind a secure screen
and the pressure and tightness of the material around my head and face was
like being squeezed constantly."[19] Another Aspie writer, ten-year-old Kenneth
Hall, said, "I love my sleeping bag so much because of how nice it feels against
my skin. I love to get right inside it and draw it tight. It is a lovely soft feel. . . .
Sometimes I curl up happily in my sleeping bag in all parts of the house, even
the kitchen floor! . . . I even thought of [becoming] a sleeping bag tester."[20]
The most innovative enclosure, however, may have been the one constructed
by British author Daniel Tammet: "I spent hours painstakingly stacking
coins, one atop another, until they resembled shining, trembling towers each
up to several feet in height. . . . Sometimes I would build several piles of equal
height around me in the shape of a circle and sit in the middle surrounded on
all sides by them, feeling calm and secure inside."[21]

As these recollections reveal, Aspies prefer deep pressure to a light feathery touch, or as Rudy Simone remarks, "We only like tight hugs . . . so we'll
avoid polite hugs people seem to enjoy giving each other."[22] Another critical factor is that the pressure must be self-administered because, as Grandin
pointed out, "it is much easier for a person with autism to tolerate touch if
he or she initiates it."[23] This makes sense when you recollect that you cannot
tickle yourself. You might not be able to stand someone else touching your
feet, but when you try to tickle them yourself, your nervous system dampens
the experience in such a way that it is bearable.[24]

These accounts make it poignantly clear that Aspies, and at least some people with HFA, crave warmth, security, and comforting contact just as much as anyone else, but that they express this need in unusual ways because of their extraordinary sensory systems. As the previous chapter showed, stimming makes sense when viewed as a soothing behavior that fulfills an ancient need for contact comfort—in other words, as a manifestation of the second evo-devo trend.

Overly Sensitive or Not Enough?

An unusual, almost paradoxical, aspect of AS and HFA is that affected individuals often have some senses that are extremely keen but others that are relatively underdeveloped. Clinicians have shown that the close-up senses of taste, smell, and touch sometimes seem impaired in autism, whereas the distant senses of hearing and vision appear to be enhanced. For example, one remarkable study compared vision in a mixed group of HFA and AS people with a group of typically developed individuals. Intriguingly, the autists had vastly better vision than the others.[25] The average vision for the autists was 20:7, which means they could see details at 20 feet that a person with average vision sees from 7 feet. According to the researchers, not even birds of prey have such acute vision! Interestingly, the authors speculated that the enhanced visual acuity of autists might be related to their well-known abilities for detecting visual details. Similar results have also been reported for hearing. Thus, a mixed group of people with either HFA or AS performed significantly better than neurotypical individuals in perceiving auditory pitches, consistent with earlier findings that they were also superior at discriminating certain melodies because of heightened pitch perception.[26]

The fact that vision and hearing are so enhanced in persons with AS or HFA probably accounts for why these senses are frequently bothersome to them.[27] Think of being in a restaurant where the supposedly "background" music is so loud and distracting that it is impossible to carry on a conversation; this is just a glimpse of what people with AS regularly experience. Like Temple Grandin, who describes being virtually deluged with sound, many Aspies consciously shut out intolerable sound and light by stimming or entering a state of flow. The aversive stimuli may be tolerated, however, if the autistic person is in control of them. Aspie author John Elder Robinson, who remembers growing up "in a world of sensory overload," explains that despite his sensitivity to light and sound, "I could make noise or flash lights as loud or bright as I wanted, with no problems. As long as I was in control, my own light and sound never bothered me. I could shriek at the top of my lungs all day and feel fine, while everyone within a hundred yards wanted to throttle me. But if someone else made half that noise or flashed a light at *me*, I went nuts."[28] Rudy Simone likewise observes, "A lot of these triggers . . . exist only in relation to our lack of control over them. If I'm running the lawnmower and making a racket, that is okay, but if my neighbor is, that's a different story."[29]

Contrary to their hypersensitivity to visual and auditory stimuli, Aspies' sensitivity to heat and cold, to touch, and to the experience of pain may be reduced. It is thus quite common for people with AS to have unusual responses to temperature, for example, a preference for wearing warm garments in hot weather or very little clothing when it is freezing. (Just try to get Eve out of her tights and jacket on a hot and humid Tallahassee afternoon!) Aspie children are also more likely to be relatively insensitive to sharp and burning pain on their skin.[30] Because she was less likely to pick up on painful cues, Liane Holliday Willey recollects, "I earned the nickname 'Miss Accident Waiting to Happen' because I'd come home with swollen ankles and skinned knees and knots on my head. I put a hot iron on my wrist when I was a kid just to see what it would do."[31] (See Eve's comments at the end of this chapter about injuries from tripping and slipping.)

For Aspies, reduced sensitivity to pain may, or may not, be more than skin deep. Heavy tactile experiences like getting clobbered ankles and skinned knees did not seem to bother Willey, a characteristic parents frequently observe in their autistic children. On the other hand, faint touches can be excruciating, which may be difficult for a neurotypical person to comprehend.[32] As Aspie author Clare Sainsbury puts it,

> Sometimes sensitivities may only be to certain ranges of
> stimuli—e.g. certain types of touch (often light, feathery touches). . . .
> This can cause problems of misinterpretation—when a child wails
> "she hit me!" after . . . a classmate accidentally brushes against them
> in the corridor, they are likely to be seen as lying or exaggerating,
> and it is hard for others to understand that they may be stating their
> perceptions accurately, while it may be equally hard for the child with
> Asperger's to understand that the hurt was not intentional.[33]

At first blush, this makes no sense. How can someone be indifferent about serious wounds but cringe at the touch of a feather? One possibility is that this is due to unusual wiring in some of the receptors that send impulses to the spinal cord and brain about particular sensations from indentations of the skin, deformation of hair, vibrations, changes in skin temperature, stretching of muscles, or destruction of tissue.[34] Although sensations from a feathery touch are rarely, if ever, perceived by TD individuals as painful, impulses from the receptors for light touch may contribute to such a perception in autists.[35] The perception of stimuli as more or less painful by people with AS and HFA are also likely influenced by unusual processing in the specific parts of their brains that participate in interpreting such stimuli (as they do in neurotypicals), for example in the insula[36] and the thalamus.[37] As these observations suggest, some of the unusual behaviors that are common in AS, such as a dislike of foods with certain textures or a desire to wear heavy clothing in hot climates, are probably responses to a fundamental, if underappreciated, feature of Aspie and HFA nervous systems—namely, unusual sensory processing.

As the Autistic World Turns

Most people think of the brain as having five basic senses: hearing, sight, touch, smell, and taste. As we have seen, though, this is an oversimplification. There are other "forgotten" senses that are very important for understanding AS, including one that contributes to an individual's perception of his or her position in space, which depends on the vestibular system in the inner ears.[38] This system is regulated by the swishing of fluids in three semicircular canals that inform the brain about the head's rotation and orientation relative to gravity.[39] Additionally, two bulges near the canals contain tiny crystals that are sensitive to acceleration as the head moves up and down, for example when one rides in an elevator. As a person moves, signals from the vestibular system stimulate reflexes in the neck and spine that maintain stability of the head and body, helping to provide postural tone and balance. With time, the vestibular system can adjust to new circumstances, which is why sailors develop the ability to maintain their balance in rough seas (get their "sea legs"), why astronauts are able to regain their balance after returning to Earth's gravitational forces, and why Eve is able to recover from the "airplane head" she gets from long flights (described below).

Suppression of jittery eye movements by visually fixating on a distant object when the vestibular system is being overstimulated prevents the disorientation, loss of balance, staggering, motion sickness, nausea, or vomiting that would otherwise occur.[40] Remarkably, several studies show that the vestibular systems of autistic children function atypically. For one thing, these youngsters are significantly better than other children at suppressing jittery eye responses when being spun in a rotating chair.[41] Other research reveals that stimulating the vestibular system in sleeping children by rocking their beds causes an increase in the rapid-eye movements of neurotypical, but not autistic, children.[42] These findings may have something to do with why many autistic children are able to spin, rotate, and twirl without becoming dizzy or nauseous.

In fact, many people with AS and HFA seem not merely immune to motion sickness from being spun rapidly—they *like* the sensation. Daniel Tammet writes that when he was a child, "there was . . . a merry-go-round, and I sat in the middle of it as my parents stood on either side and slowly moved it along. As the merry-go-round spun again and again I closed my eyes and smiled. It made me feel good."[43] As a child, Temple Grandin remembers, she also "loved to spin, and I seldom got dizzy. When I stopped spinning, I enjoyed the sensation of watching the room spin."[44] Significantly, Grandin notes, this stimming behavior made her feel "powerful, in control of things. After all, I could make a whole room turn around."[45] More than that, Grandin and other Aspies may use spinning to relax. Grandin recalls the range of feelings she experienced on a carnival ride when she was an adolescent:

> The rotor picked up speed and the motor sounded like a giant's hum. The colors of the blue sky, the white clouds, the yellow sun blended together like a spinning top. . . . Fear tasted bitter in my mouth. . . . With a creak of the hinges the floor opened to the ground below but now my senses were so overwhelmed with stimulation that I didn't react with anxiety or fear. I felt only the sensation of comfort and relaxation. After the ride I was at ease with myself for the first time in a long, long time. Again and again I rode in the barrel, savoring first the over-stimulation of my senses and then the quiet surrender of my panicky, anxious nervous system.[46]

In addition to enjoying the effect of being spun or of spinning themselves around, Aspies frequently like to spin objects, such as tops or ceiling fans, and watch them spinning. Daniel Tammet observes that "one of my favorite pursuits was taking a coin and spinning it on the floor and watching it as it spun round and round. I would do this over and over, never seeming to get bored."[47] Grandin noted that spinning objects put her into an enjoyable altered state: "I enjoyed . . . spinning coins or lids round and round and round. Intensely preoccupied with movement of the spinning coin or lid, I saw nothing. . . . no sound intruded on my fixation. It was as if I were deaf."[48]

Although the vestibular system is crucial for perceiving one's position in space, it is aided by another "forgotten" sense called proprioception, in which signals from special receptors in the joints, muscles, and tendons continually inform the brain of where the parts of the body are relative to each other and how they are moving.[49] You can experience proprioception in action by closing your eyes and touching the end of an index finger to the tip of your nose. This sense of proprioception helps us stay grounded, maintain posture, and move about in the world. These tasks become difficult if the sense is temporarily lost, as you may have experienced by trying to walk on a leg that has gone to sleep. When proprioception is not working properly, a child is likely to have clumsy, poorly coordinated movements. Consequently, he or she may seek to regain a sense of balance by a variety of means—stretching, applying pressure by squeezing into tight spaces, crossing or twisting limbs around each other, clapping or flapping the hands, or kicking heavy objects. In other words, these children adopt postures that help "provide essential extra tactile and proprioceptive information to the brain about where the child's limbs are in space, and they also confirm for the child that their body is securely 'fixed' and not moving or floating around."[50] (As she will describe, Eve fears being adrift in water.)

Though they have been little discussed, the vestibular and proprioceptive systems are *hugely* important for understanding not only the sensory aspects of autism, but also the unusual motor behaviors that are still being used to diagnose it. For example, the repetitive mannerisms of autistic children strongly suggest that they actively seek stimulation from these two systems: "They whirl themselves around and around, repetitively rock and sway back and forth, or roll their heads from side to side. The repetitive hand-flapping also provides proprioceptive input."[51] Even more intriguing, "some of these [movement] patterns can at times be elicited by rapidly spinning a child's top in front of the patient."[52]

Spinning and rotating objects seems to be a way that autistic children learn. For example, one study showed that, by as early as twelve months of age, infants who were later diagnosed with autism spent significantly more time rotating and spinning things than their neurotypical peers. The infants also spent much longer exploring objects from odd angles or out of the corners of their eyes (using peripheral vision), often while squinting and blinking.[53] The authors concluded that autistic children rely on internal position and manipulation cues more than other kinds of perception, and that they comprehend their environments by focusing more on sensory feedback from their own movements than neurotypical children do.[54] Another study found that "the greatest discrepancy between the AS subjects and TD controls occurred when proprioceptive influences were central to . . . performance" and (consistent with this chapter's focus) that the clumsiness frequently observed in AS may be from deficits in sensory rather than in motor processing.[55]

As we have seen, the "forgotten" vestibular and proprioceptive senses are not only crucial for interpreting core behaviors of AS such as stimming but also, literally, generated within the "core" joints, muscles, ligaments and inner ears of the body. Besides being important for understanding the clinical deficits that are at the very heart of AS, these senses help explain how the deeply engrained need for contact comfort that first emerged in prehistoric infants is manifested in youngsters with AS and HFA. In this context, the poignant observations of Irenäus Eibl-Eibesfeldt, the founder of the field of human ethology, are particularly apt:

> The rocking, jostling, and lifting of infants, which mothers enjoy so much, accommodates the infantile need for vestibular stimulation. An excited infant can be calmed by rocking it. In kin-based societies infants spend most of their time being carried about by the mother or another person. Vestibular stimuli communicate to the infant that he is not alone. Hospitalized children isolated from this stimulus often develop movement stereotypes, like rocking and self-patting, that serves as self-stimulation.[56]

It's not just the motor systems of Aspies that are affected by their unusual vestibular and proprioceptive senses. As we will see in the next chapter, the failure to develop a spontaneous subconscious sense of movement and spatial orientation also prevents Aspie babies from acquiring a "feel" for the experiences of others (empathy), a deficit that profoundly affects their cognitive development.

Eve's observations provide a personal glimpse into the ways that sensory experiences form an important but underappreciated core of AS and HFA. Apprehensions and physical difficulties are a big part of Eve's life, and these can often be traced to deficits in her sense of balance and lack of intuition about where the parts of her body are relative to one another and how they move. Her other senses show a combination of hypersensitivity (e.g., vision, hearing) and hyposensitivity (e.g., to hot temperatures and spicy food) that is common for AS. And like other Aspies, Eve has come up with remarkable ways to compensate for her atypical vestibular and proprioceptive senses, such as hanging on to family members as a navigational aid and distracting herself from her fears with books. Consistent with Hans Asperger's observations over seventy years ago, Eve spends a good deal of time in her own head rather than socializing. The fascinating imbalance between social awareness and intellectual capacity that typifies AS, and what is behind it, are subjects of the next chapter.

Many reports fail to convey the wallop that the combined senses have on the day-to-day lives of Aspies. For that, we will now turn to Eve. Like the people mentioned above, she remembers having enjoyed twirling in swing sets and going on roundabouts as a child ("a real rush"), and retains acute senses of hearing and vision as a young adult. Like other Aspies, she is superb at shutting out or minimizing aversive stimuli by "getting in the zone." As Eve describes her sensory experiences, it is clear that her greatest fears and physical difficulties are associated with balance and height issues and a lack of awareness about where parts of her body are located and how they are moving. Eve's description of the role that sensations play in her life gives a good idea of what it is like to be an Aspie.

EVE: SENSATIONS AND FEARS

I am often overwhelmed by sensations or feelings. For instance, after I get off an airplane I continue to feel as if I am bouncing on air for a day or so. It feels like my head and shoulders are jostling about like they were on the plane, even though, on the plane and off, those parts of my body are actually still. I usually think it has something to do with jet lag or the change in time zones. I just try to sleep or relax and ignore whatever tension there is in my body and spine until my "airplane head" goes away.

Many of the sensations that bother me have to do with my fears of water, heights, and falling. It is the feel of weightlessness in water that gets to me. As soon as I get into a pool, it feels as though I'm floating up and away from the solid ground, which isn't very nice. I would liken it to constantly being about to fall over the edge of a cliff and trying to keep yourself from doing it. Even when I was older, I preferred using the ultra-shallow kids' pool to the adults' pool. It, at least, was somewhat warm and I didn't feel as though I was going to lose my balance when I went in. That is the key to my fears where water is concerned. I don't want to lose my balance in the water because it could lead to drowning. I also don't like getting water in my eyes, which are incredibly sensitive to physical sensation.

I remember going to the Corning Glass Museum as a kid, and that the floors were all open enough that you could see each one of them from the one above. It did not do wonders for my fear of heights and always made me move extremely slowly between the exhibits. Of course, the shop on the ground floor had so much color and beauty to it that I didn't find it as boring as I usually find shopping. It had a really nice gumball machine that had this entire procession thing going on before it let the gumball out. It went down this elevator-like thing, then rolled down this xylophone-like thing and did a bunch of other stuff before letting the gumball slide down a slide to the part of the machine where you could retrieve it. All in all, I thought the glass museum was quite fun as long as I could avoid looking over the balconies and concentrate on the exhibits.

When I am facing something I really fear, such as going down a steep hill, or using an escalator without help, I can have a meltdown. I tend to do better if I either have a book to concentrate on, or someone to hold on to that helps to ground me. My sisters, Helen and Judith, or my dad can get annoyed with me for doing this every time we go up or down an escalator, but I am scared stiff of heights enough to throw fits or just start breathing in short bursts. In either case, my voice keeps rising until I end up screaming or shouting, sometimes without even knowing it, until a relative or friend tells me to lower my volume. I also end up freezing in place when I'm scared. It used to be worse. Now I just have to be holding on to someone I trust when the fear strikes.

For example, my mom and my stepdad, Michael, once took my sisters and me to a nature center that had a tree house that went up seven stories. To reach this monstrosity, my sisters and I had to walk across a wooden path inlayed into the forest floor. I was fine with it despite my usual reservations about anything remotely resembling a bridge, especially one that would end up moving while I was walking across. It's a combination of being up high and the fact that I am scared of water. Anyway, my sisters went into the tree house and began looking at the exhibits about nature, and I have to admit some of them were interesting. Helen and Judith went right up to the other stories, but I resolved to stay on the ground floor. I wasn't scared until I looked out the window and saw that the tree house overlooked a huge ravine. When the wind moved the branches of a tree right on the edge of the ravine, I thought that I felt the tree house move. Because I was already nervous, that put me over the edge and I sat on the floor shaking and too scared to move. To put it mildly, I freaked, and Michael had to carry me, crying and screaming, out of the tree house.

On another visit to a museum, one of the more overwhelming exhibits was the animatronics of the dinosaurs, which scared me because the roars sounded so loud. But, then, some sounds can be overwhelmingly loud while, at other times, I find myself having to turn the volume up higher to hear what I am supposed to be listening to. Maybe it is just because if I lose my concentration on something the awareness of it can fade away or just become a part of the background. Of course, it could be said that this fading of sound was another way to deal with a fear of things being too overwhelming. I also tend to get so caught up in the things that interest me that I forget all else, which is a bad thing for my schoolwork. I tend to get dizzy or tired when I disengage from my interests.

When I was a child I sometimes ended up with injuries to my legs because I just didn't see what was happening around me and tripped on something or slipped. There were also times when I tumbled down the stairs at my house. It's not that I'm clumsy per se, it's more like I don't have a sense of my physical body. When I walk in public with my family, they have to constantly drag me out of the way of people because I'm so intent on concentrating on something else. On occasion, I try to remedy this by physically wrapping my hands around the arm of whatever family member is nearest to me and holding on to them. It seems to keep me in line with them and out of the way of other people because the rest of my family actually have a sense of where they are physically in relation to other people.

I have trouble with inappropriate facial expressions because I am never really sure what my face looks like from feeling alone. In other words, I don't sense my own facial expressions. I had trouble with my speech as a child because I didn't seem to know when different parts of my face moved and that included my tongue. The reason for the eye contact avoidance is simple. It just seems that half the time I am not aware of what my face is doing, and that includes my eyes, which is why I find it disconcerting when I end up staring at my nose from one side, and looking at something else from the other.

Not all of my sensory experiences are related to my fears or lack of a sense of my own body. I enjoy touching the hair of people around me. In particular, I like to touch curly hair or just hair with a fine soft texture. Or maybe I just like the feel of the skin underneath the hair because I also like to rub the scalp of people who are bald or tap it. Though I really only do that last thing with family members. I call it playing the head bongos and, to me, it is fun though it annoys my family members. So, yeah, I do like the feel of hair, but it is more about keeping some of the habits from when I was a kid. Other things that I like to touch include cat fur, and the fur on some other animals. As long as it is a soft texture and not a rough or smooth one like sand. Those just feel annoying to me.

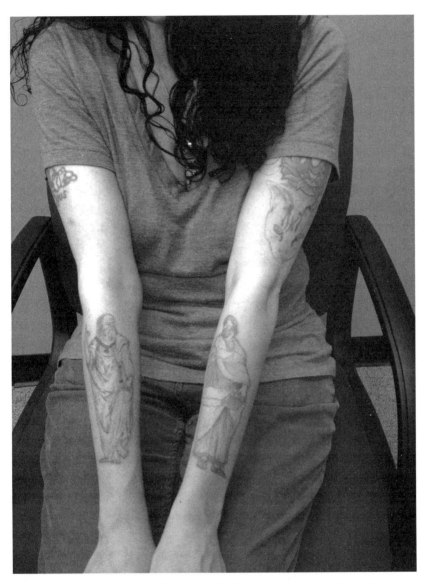

Figure 2.1. Eve's enchantment with texture. "When I was at Grandma's working on this book," she says, "I met some of her philosopher friends, including the graduate student in this photograph who had long curly hair, which she let me touch. She also had fascinating tattoos. I was interested in the cat she has near her shoulder and the ohm character above it. That is Plato pointing upwards on her right arm, and Aristotle on the left." Photograph taken by Karen Foulke.

Chapter 3

Social Naïveté and Advanced Cognition
An Aspie Enigma

> The personality of the children presented here lacks, above all,
> harmony between affect and intellect.
>
> —HANS ASPERGER[1]

ans Asperger was, indeed, spot on when he observed a disconnect between Aspies' moods, feelings, and emotions (affect) and their intellects. Furthermore, this enigma appears to be linked to distinct styles of social interaction in people with Asperger syndrome and high-functioning autism, as shown by University of Michigan psychiatrist Mohammad Ghaziuddin. Ghaziuddin was struck by numerous reports that people with AS, rather than being loners like many autists, attempt to interact socially but without understanding social rules; for example, they may ask inappropriate questions and talk at length about their favorite subjects. Ghaziuddin investigated the social interactions of 58 AS and 39 HFA males and females of various ages.[2] He found that each person fell into one of three categories: *aloof individuals*, who were indifferent in most social situations; *passive people*, who responded to questions without adding to the conversation; and *active but odd individuals*, who initiated social interactions, often with inappropriate questions, and who were frequently fidgety. Even after factors such as IQ were taken into account, a majority of people with HFA were categorized as aloof or passive, whereas most Aspies were deemed to be active but odd. These results confirmed that "persons with AS often have a distinct style of social interaction."[3]

Two types of social processing are of particular interest in studies of AS and HFA: (1) the ability to infer the beliefs, plans, and desires of others, called theory of mind (ToM), and (2) an associated ability, not only to understand the feelings of others at a cognitive level but also to *feel* them (empathy).[4] The term "theory of mind" was coined more than thirty-five years ago in association with studies to determine if chimpanzees who viewed videotapes of caged human actors struggling to obtain inaccessible bananas could understand the actors' intentions and indicate a solution by selecting an appropriate photograph from alternatives, such as one of the human stepping on a box to reach the fruit.[5] Chimpanzees, who had long been known to be capable of solving similar problems for themselves, proved up to the task, which indicated that they were able to impute mental states, not only to themselves, but also to others—in other words, they had ToM or, as some prefer to characterize it, an ability for "mindreading" or "mentalizing."

Since these pioneering studies, research on ToM has become a major area of focus, not only in primatology, but also for child psychology and studies of various clinical conditions including autism. Experiments show that TD infants as young as one to two years old demonstrate "implicit" mindreading capabilities by spontaneously pointing to objects in response to observing others searching for them.[6] Relatively easy "false-belief" laboratory tests are routinely used to explore ToM in somewhat older neurotypical and autistic youngsters, such as the famous Sally Anne test, which involves two dolls named Sally and Anne.[7] In brief, Sally places a marble in a basket in Anne's presence then leaves the room. Anne then removes and hides the marble in a box. When Sally returns, the experimenter asks the child, "Where will Sally look for her marble?" If he points to the basket, he passes the belief question. At a slightly more difficult level, the experimenter might ask, "Where does Anne think that Sally will look for her marble?" Most TD children are able to pass false-belief ToM tests between the ages of three and five. Children with HFA, on the other hand, are often severely delayed well beyond this age.[8] Not surprisingly, the key factor that predicts the level of performance for both autistic and TD children is the degree of development of their grammatical skills, perhaps (at least partly) because such skills facilitate preschoolers' participation in "family conversations about thoughts and feelings."[9]

Very young Aspies, on the other hand, may[10] or may not[11] display "mindblindness" on false-belief tests. As a rule, however, AS youngsters perform much better than those with HFA on this type of test, reaching the level of neurotypical children by the time they are eight years old, although it is "conceivable that typically developing preschoolers may begin to master false belief somewhat earlier than those with AS."[12] This raises the question of whether AS youngsters are simply late bloomers when it comes to ToM, as they are for achieving milestones like crawling, walking, etc. (the first evo-devo trend discussed in chapter 1), or whether ToM in Aspies and neurotypical individuals differs in more subtle ways that aren't revealed by relatively easy false-belief tests taken under relaxed laboratory conditions. After all, Aspies' well-known "active but odd" social interactions suggest that, try as they may, Aspies are not particularly adept at social interactions that require "getting" other minds.

Intriguing research of Norwegian psychologist Nils Kaland and his colleagues has done a great deal to answer this question. The authors assessed the ability to grasp advanced ToM nuances embedded in a variety of complicated but realistic *Stories from Everyday Life* in Aspie and neurotypical children and adolescents.[13] There were thirteen types of stories, each of which probed comprehension of social communication entailing one of the following: a lie, white lie, figure of speech, misunderstanding, double bluff, irony, persuasion, contrary emotions, forgetting, jealousy, intentions, empathy, and social blunders. After reading or hearing each story, the individuals answered ten to fifteen questions, which included one about a particular physical aspect of the vignette and two later questions about the mental states of the story's characters. For example, a story used to examine the ability to grasp irony was a vignette about two brothers who had been asked by their mother to clean their rooms: "Tom, the youngest of the brothers, is always making a mess, and his room is usually very untidy. His mother often complains about the mess. Adrian seldom has to hear such remarks, but his mother says that he should now and then help his father tidying the villa garden."[14]

After the two boys had been tidying for a while, their mother asks if they will soon finish, to which Adrian responds that he already has. However (goes the story): "Tom hasn't begun to tidy up at all! Adrian's mother asks if he can look in Tom's room to check if he has tidied up. Adrian opens the door to Tom's room, peers in, and sees that the room appears as it normally does. He shouts to his mother: 'Mother, Tom has as usual done a splendid job tidying up!'"[15] In this case, the physical question was "How does Adrian's room look?" which required making an inference about a physical state. The person taking the test was then asked, "Is it true what he [Adrian] says?" and then "Why does Adrian say this?" In order to correctly answer, the listener had to understand that Adrian wasn't telling the truth and that his comment was sarcastic or ironic.

Aspies found it much more difficult to make inferences about mental processes than physical states in these stories, and many were slow and long-winded or gave idiosyncratic answers compared to TD individuals. They also tended to interpret events literally, which contributed to their particular difficulty in grasping mental states in certain types of stories, such as those about contrary (conflicting) emotions, figures of speech, and empathy. For example, Aspies tended to interpret "castle in the air" as an extremely high building. Interestingly, they were much better at understanding some kinds of stories, such as those about lies and jealousy, but they were still behind TD individuals in their level of comprehension. The bottom line is that Aspies, indeed, have difficulty applying their cognitive knowledge about ToM to complex social situations in real life settings that require rapidly processing subtle cues, although they are usually better at it than people with HFA.[16]

Researchers use various definitions for the second type of social processing that is widely studied in autism—empathy. Some believe empathy has a cognitive component involving recognizing and sympathizing with the feelings of others, and perhaps even wanting to help them, and that this overlaps with another strictly emotional (affective) component.[17] Others view the cognitive component as "affective ToM" or, more simply, thinking about feelings, as distinct from emotional empathy, which entails the observer actually *feeling* (mirroring) the emotions of others.[18] Whichever view one takes (which is a matter of taste), it is important to understand that in TD individuals, three different but overlapping neurological networks are active in cognitive ToM, affective ToM, and emotional empathy, and that balanced activity across all three networks supports appropriate social behavior.[19] As described below, these networks are wired somewhat atypically in people with AS, and activity across them differs from that of neurotypical individuals.[20]

It is well known that Aspies are challenged when it comes to empathy, which, no doubt, contributes to their difficulties connecting socially despite their desire to do so.[21] Research that ferrets out various other aspects related to "emotional intelligence" may shed some light on Aspies' difficulties with empathy. For example, one study of young adult Aspies found that their ability to perceive and think about emotions in laboratory tests was not only intact but also actually better than that of TD individuals, presumably because they used verbal skills and step-by-step logical thinking to work things out.[22] These same individuals, however, rated low on affective traits such as optimism, self-awareness, and self-esteem, which are relevant for processing emotional knowledge at a gut level in real-life social interactions. In the words of the authors, "Even though actual knowledge and skills in emotional situations seems intact, performance in real life situations remains problematic."[23] What this boils down to is that, smart as they may be, Aspies do not "naturally seek social meaning in the environment."[24]

This is certainly the case for Eve, who writes below, "I critically lack what most people call empathy." Eve's lack of empathy makes sense, however, when one recalls that she also lacks normal proprioception, as she noted in chapter 2 ("I don't have a sense of my physical body. . . . I am never really sure what my face looks like from feeling alone. In other words, I don't sense my own facial expressions"). Such proprioception is essential for developing ToM and empathy in young infants, according to American psychologist Andrew Meltzoff and his colleagues. Meltzoff reasons that very young human babies have never seen their own faces because "there are no mirrors in the womb," and that, although they cannot observe their own facial movements, they are still able to *feel* them.[25] Thus when newborns imitate adult acts such as lip and tongue movements, they associate the visual observation of the adult with the feeling of performing the act themselves.[26] This association is stored in memory, which allows infants to infer the experiences of others. An important implication is that infants' perception and production of motor behaviors are intertwined, and the same kind of '*that*-looks-like-*this*-feels' neurological coding (which takes place during the early brain spurt) helps them infer visual, tactile, and motor experiences of others.[27] As is clear from chapter 2, Eve and other Aspies tend to have problems with the sensory end of this equation, which may help account for the challenges they face when it comes to processing ToM spontaneously and experiencing empathy.

Some evolutionary biologists believe that advanced ToM is a highly evolved human ability, and a few have even suggested that its manifestation in the form of self-interested "Machiavellian intelligence" was a direct target of natural selection.[28] A less ominous theory is that an ability to feel, or mirror, the emotions of others may have provided our ancestors with an "emotional stake" in others' welfare, contributing to the likelihood that they and their relatives would do well in the evolutionary sweepstakes.[29] Although it is true that ToM is relatively sophisticated in humans, it is by no means unique in them. (Recall that studies of mentalizing chimpanzees inspired the research on human ToM.) Monkeys and apes tend to be extremely social and engage in a certain amount of ToM and empathy, but these behaviors do not come close to the complexity seen in the social machinations of humans. This is not surprising, given that the linguistic skills, particularly grammatical ones that are unique in humans, are positively associated with the development of ToM in both autistic and neurotypical children.[30] It seems reasonable, therefore, to suggest that the emergence of human-level mindreading and empathy depended on the distributed neural networks that evolved in conjunction with language—networks that eventually facilitated the emergence of humankind's advanced scientific and creative abilities.[31]

Basic Anatomy of the Human Brain

Before we can explore features of the brain that are associated with AS and HFA, we need to review some basic information about the neurological underpinnings of thinking. The simplest types of thoughts begin with stimulation of parts of the brain that process incoming sensory information from the ears, eyes, skin, muscles, and joints—the so-called primary sensory areas. This information is usually received on the side of the brain that is opposite to where it originated in the body. The primary areas that sense touch, hearing, and seeing are located, respectively, in the parietal, temporal, and occipital lobes, as illustrated here for the left side of the brain (fig. 3.1).

Movement of the body is controlled by the primary motor cortex in the frontal lobe, which is directly in front of and parallel to the primary sensory cortex in the parietal lobe that processes sensations from the body's surface. As shown in figure 3.1, these sensory and motor representations are mapped upside-down on the brain, so that the tongue is below the head, which is below the hands, and so on. Notice that only the right hand is mapped on the left hemisphere, which is because body parts are usually represented on the opposite side of the brain. (The representations for the right foot are not visible in the figure because they are mapped on the inside surface of the left hemisphere.) After newly arrived signals are integrated in the first-stop (or primary) regions for sensing the body, hearing, and seeing, the results are sent to nearby association areas for processing within a wider context that incorporates memory as well as information from the other senses. As just

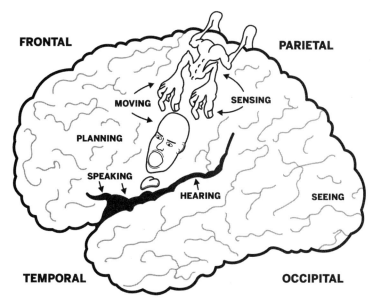

Figure 3.1. Basic floorplan of the human brain. Simplified schematic of the left side of a human cerebrum. Four lobes and regions associated with certain functions are labeled. Notice that the cartoon map (called a homunculus) shows how the brain is organized for processing sensory information from the body in the parietal lobe and movements of various parts of the body in the frontal lobe. Speaking is primarily a function of the left side of the brain. The other labeled activities are facilitated by both sides of the brain. Illustration modified from Falk, *Fossil Chronicles*.

one example, if someone were to close the palm of your right hand around an unknown small, cold, irregularly shaped object, the association area in your left parietal lobe would synthesize various aspects of the experience and, voilà, you would identify the thing as a key.

One of the main activities of the frontal lobes is planning and carrying out movements that are facilitated (in part) by the primary motor cortex. This is done through an "executive control network" that is dispersed mostly on the outside surface of the frontal and parietal lobes (fig. 3.2). The executive network influences sensory and motor functions and uses working memory to keep goals in mind until the frontal lobe carries them out.[32] The network receives crucial input from another system, called the "salience network," which includes the front part of the insula (a lobe buried deep within the

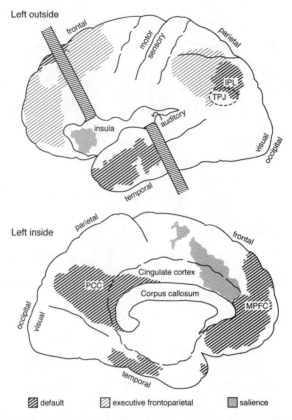

Figure 3.2. Lobes and key networks of the human brain. *Top:* outside surface of the left hemisphere, with the temporal and frontal lobes separated by tongs to reveal the normally hidden insula; *bottom:* inside surface of the left hemisphere. The lobes of the brain are labeled, along with the basic kinds of primary sensory information they process (visual, auditory, sensory representations from the body [fig. 3.1]) and the primary motor region. Three important network systems are shaded as indicated. The executive part of the frontal lobe is important for making decisions and carrying them out. Among other functions of the default system, the inside part of the temporal lobe conjures up previous memories, and the medial prefrontal cortex (MPFC) of the frontal lobe uses this information when thinking about oneself. The posterior cingulate cortex (PCC) integrates information from the first two regions and is involved in personal memories, visual imagery, and thinking about one's future. The default system also includes outside parts of the brain, most notably the inferior (lower) part of the parietal lobe (IPL), which includes the back of the temporo-parietal junction (TPJ). The salience system recognizes urgent (salient) information by integrating gut-level sensations with information received from various parts of the brain that are concerned with emotions, monitoring conflict, and anticipating rewards. It provides an important interface between affective and cognitive processing. Based on Buckner, Andrews-Hanna, and Schacter, "The Brain's Default Network"; Bzdok et al., "The Temporo-Parietal Junction"; Fox et al., "Anticorrelated Functional Networks"; and Seeley et al., "Dissociable Intrinsic Connectivity Networks." Prepared by Sandra Vreeland.

brain) as well as a small part of the cingulate cortex (and a bit above that), which is visible on the inner surface of each hemisphere (see fig. 3.2).

The salience network communicates with other deep structures near the middle of the brain that are part of the limbic system, which processes information related to smells, emotions, and memories. The salience system recognizes urgent (salient) information that helps the frontal lobe (and other regions) react to a variety of external and internal stimuli.[33] It does so by integrating gut-level (autonomic) sensations with information received from various parts of the brain that are concerned with emotions, monitoring conflict, and anticipating rewards. Because it influences the frontoparietal executive system, the salience network is recognized as an important interface between affective and cognitive processing. A third network illustrated in figure 3.2, the default network, is typically active when an individual is not paying much attention to the outside world. We shall return to these networks when we discuss the brains of autistic individuals.

During the course of hominin brain evolution, the size of the parts of the brain that integrate information from various senses (association cortices) enlarged, the number of subareas of the brain increased, and the networks that are active during different kinds of mental activities became more complicated.[34] As discussed in chapter 1, these changes likely piggybacked on the emergence and evolution of language. In the vast majority of people, language (including reading and writing) is processed mostly in the left hemisphere—sensory aspects in back and speech (motor output) in front (fig. 3.1). Language, like many functions, is thus said to be lateralized in the brain (fig. 3.3). Hands, especially the right one, which is controlled by the left "dominant" frontal lobe, wave around and help get the message out when people talk. A large part of the emotional content of language, however, is conveyed by tone of voice, which is understood and produced mostly on the right side of the brain. The cognitive and emotional components of language are processed together through fibers that cross between the two hemispheres via the largest pathway of white-matter connections in the brain, the corpus callosum, which is located deep within the brain (fig. 3.2, bottom; fig. 3.3).[35]

As our ancestors evolved linguistically, their brains became increasingly lateralized compared to those of other primates.[36] In addition to language, the left hemisphere eventually evolved specializations in other faculties that entail rapid processing of tiny sequential signals, including reading, writing,

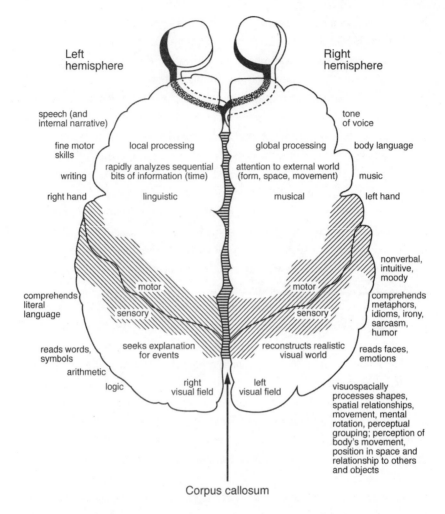

Figure 3.3. Brain lateralization viewed from the top of the brain. This imaginary slice through the brain, which shows the eyes at the top and the visual cortex at the bottom, illustrates the right-left differences that are found in the brains of most neurotypical individuals. Although the named functions are facilitated more on one side of the brain, it is important to note that they are usually influenced by input from the other hemisphere through fibers that cross between the two sides of the brain via the corpus callosum (arrow). A small minority of TD people are reversed in some, but not necessarily all, of these asymmetries. The information in this illustration comes from numerous sources discussed in the book. Note the distinct "personalities" of the precise and analytical left versus the holistic and intuitive right hemisphere. Prepared by Sandra Vreeland.

and arithmetic.[37] Because of its ability to analyze tiny signals as they unfold in time, this side of the brain plays a large part in logical, analytical, and systematic thinking. In other words, it is the left side of your brain that brings out the scientist in you.

Not so the other side of the brain, which is intuitive rather than verbal, and focuses on interpreting the external visual world rather than internal details.[38] Besides tone of voice, the right hemisphere of neurotypical individuals is more involved than the left hemisphere in processing music, humor, sarcasm, metaphors, irony, and idioms (fig. 3.3).[39] In right-handed people, this side of the brain also processes emotions more than the left[40] and is better at reading faces.[41] The right hemisphere pays greater attention to visuospatial information and is more interested in grasping the big picture than in understanding little bits of sequential information the way the left hemisphere does.[42] Curiously, part of the right parietal lobe is more involved than its counterpart on the left in processing information about the interrelationship of one's own body parts and how they move and relate to surrounding space and objects.[43]

Neurotypical Thinking

Thanks largely to functional magnetic resonance imaging (fMRI), scientists have shown that thinking is an incredibly complex and dynamic enterprise, with different parts of the brain "lighting up" (activating) and deactivating depending on what is going on in an individual's head at any given moment.[44] Remarkably, when people are relaxed with their eyes closed and thinking about nothing in particular, for instance when resting, sleeping, or under anesthesia,[45] some parts of their brains share spontaneous, synchronized patterns, almost as if these regions were humming the same tune or engaged in crosstalk.[46] Approximately twenty such "resting state" networks (tunes) have been identified, and each of these contains many subnetworks.[47] When individuals are not resting, these networks facilitate specific "task-positive" sensory, motor, or cognitive processes like those described above for the executive control and salience networks.[48] In other words, "the resting brain's functional dynamics are fully utilizing the set of functional networks as exhibited by the brain over its range of possible tasks."[49]

Although most of the resting-state networks become more activated during task-positive thinking that engages their functional specialties, there is one remarkable exception. One of the twenty primary resting-state networks actually deactivates when individuals turn their attention to task-positive cognitive activities. Because this system is the only one that appears to be active only when the minds of resting people have nothing better or more urgent to do, it is called the default network (see fig. 3.2).[50] It would be a mistake, however, to think that people must have blank minds for the default network to be activated.[51] When their minds wander, spontaneous internal ruminations and ideas pop up that activate various parts of this network. These thoughts may include daydreams, fantasies, reflections (memories) about the self or one's past, and inferences about the motives, intentions, and emotions of other people. In other words, the default system is active when people's thoughts turn inward, including during social cognition and ToM.[52] It's as if the other resting networks go out to play, and the default network stays indoors and ruminates.

Intriguingly, the internally focused default network is also active when one watches movies,[53] reads fiction,[54] or thinks about the future. The latter is particularly important because, as psychologists Daniel Gilbert and Timothy Wilson put it, "Alas, actually perceiving a bear is a potentially expensive way to learn about its adaptive significance, and thus evolution has provided us with a method for getting this information in advance of the encounter. When we preview and prefeel its consequences, we are soliciting advice from our ancestors."[55]

Activation of the default network under such circumstances makes perfect sense when one considers the functions of its various components, which are illustrated in figure 3.2.[56] The three main hubs of the default network are the bottom part of the medial prefrontal cortex, the posterior part of the cingulate gyrus and nearby regions, and the inferior part of the parietal lobe, including where it overlaps with the back of the temporo-parietal junction (in figure 3.2, these are, respectively, identified as MPFC, PCC, and IPL / TPJ).[57] Among other functions, the medial prefrontal cortex is activated when individuals engage in ToM and introspection,[58] whereas the posterior cingulate gyrus and nearby region is linked to remembering specific incidents from the past, including autobiographical memories, visuospatial imagery, thinking about the future, and imagining oneself in the past or future.[59]

The temporo-parietal junction (TPJ), on the other hand, performs strikingly different functions on the two sides of the brain. The left TPJ is involved in language, whereas the corresponding region on the right attends to spatial information (consistent with the above discussion about brain lateralization) but also processes two broad types of information depending on which part of it is busy.[60] The front of the right TPJ pays attention to sensory information and uses it to help the frontal lobe executive system formulate motor responses to the external world. For its part, the back of the right TPJ is involved with memory and social cognition, including ToM, social judgments, empathy, and moral decisions—in short, with default network functions. This makes sense because the back of the right TPJ overlaps with the inferior parietal lobe, which is one of the main hubs of the default system (fig. 3.2). When either end of the right TPJ is active, the other is less so, which is why the right TPJ is thought to be a possible switch between the external and internal mindsets that are associated with networks that pay attention to external stimuli and the default network, respectively.[61] The default system is of particular interest for this book because its functions are responsible for just the sorts of affective and social behaviors that are a challenge for Aspies and folks with HFA.

Functional Brain Connectivity in Combined Groups of HFA/AS Individuals

Given that social awareness and behavior tends to be atypical in autism, it is not surprising that the default network has been extensively studied and compared in autistic and neurotypical individuals. The problem, as mentioned before, is that many studies lump all kinds of autism together, which blurs features that may characterize specific groups. Nevertheless, certain interesting findings float to the top at the high-functioning/Asperger end of the spectrum. For example, when two resting-state networks were studied in groups of HFA, AS, and TD adults, one, the default network, was found to be much less active in autistic individuals as compared to neurotypicals. The other network, the dorsal attention network, which is involved in tasks like sustained attention, remembering what one intends to do (working memory), and math calculations, was active to a similar degree for autistics and neurotypicals alike.[62] The finding of reduced activation in the default network of HFA/AS adults has since been confirmed by many studies.[63] Thus the default network may be "doing something different at rest in individuals with autism."[64]

One must be cautious about generalizing from studies involving only adults, however, as shown by a study that investigated resting-state brain activity in HFA children, aged seven to twelve years old.[65] Instead of exhibiting less activity in their default networks than neurotypical children, as one might predict from the above-described study of adults, parts of the autistic children's default systems were actually hyperactive. In fact, of the ten resting-state networks examined in this study, significant differences were found in six, and in all cases activity was greater in the autistic than in the TD children. Further, the "opposite pattern of results (TD greater than ASD functional connectivity) was not observed in any brain region for any of the networks examined."[66] Apparently, resting-state neurological activity is more pronounced in many networks of HFA/Aspie children than in TD youngsters but subsides in the autistic individuals as they grow up, a phenomenon that may be related to the early brain spurts discussed in chapter 1.[67]

Hyperactivity in one resting-state system, the salience network, was especially striking in HFA/AS compared to neurotypical children. As noted, this network (fig. 3.2) integrates gut-level internal states with external sensory information in light of memory.[68] The salience network has a general role in directing attention to stimuli that are most important or relevant (i.e., salient) for guiding behavior. Significantly, the authors found that the degree of hyperactivity in this network was associated with the severity of restricted and repetitive behaviors in autistic children, and concluded that "salience network hyperconnectivity may be a distinguishing feature in children with ASD."[69] Similarly, other research showed activity in the salience network was enhanced in HFA children and adolescents when viewing pictures related to their special interests.[70]

Argentinian neuroscientist Pablo Barttfeld and his colleagues compared activity in numerous networks of HFA/AS and TD adults, including the default system.[71] Activity was measured in the networks when individuals paid attention to their body's breathing (interoceptive state) and, alternatively, when they detected unusual tones that were camouflaged by noise (exteroceptive state). Intriguingly, connectivity in the networks changed in opposite ways in autistic and neurotypical individuals when their attention shifted from the external world toward internal bodily information. In HFA/AS people, brain activity was relatively decreased during external attention but increased when attention was directed inward (especially in the salience network, the frontoparietal executive system, and, to a lesser extent, the default system—that is, the three networks shown in figure 3.2):

> When subjects are asked to focus attention to external stimuli, the associated brain connectivity network reveals sub-optimal metrics, suggesting that ASD networks are badly suited for this kind of information processing. When attention is focused on internally generated stimuli, ASD brain networks improve their metrics—even surpassing those of typicals'—suggesting that ASD networks may be better tuned for interoception. . . . It is the balance and switch between exteroceptive and interoceptive information and the importance that they are assigned that could be different in ASD.[72]

In keeping with other studies, Barttfeld found that activity in the salience network increased during inward attention in HFA/AS individuals in pace with the severity of their autism. Barttfeld's research strongly suggests that, compared to the brains of neurotypical people, the brains of HFA/AS individuals activate better organized and more efficient neurological networks when their thoughts turn to internally generated stimuli (at least those that are not of a social, emotional, or introspective nature) even though their default networks are relatively hypoactive in resting states. Thus "it is possible that ASD subjects are more likely to focus on 'internal sensations' than typical [i.e., TD] subjects"[73]

Despite the importance of these findings, studies that lump HFA and AS together must be taken with a grain of salt because, as emphasized in this book, language development differs dramatically between the two groups. What one would like are comparisons limited to Aspies and neurotypical people. Needless to say, such studies are hard to come by because of *DSM-5*'s elimination of Asperger's Disorder. Hard but, fortunately, not impossible.

The Brain and Cognition in Aspies

Sophia Mueller and her colleagues at Ludwig-Maximilians University in Munich compared the brains in adult Aspies[74] with those of neurotypical individuals using not only resting-state functional connectivity but also imaging techniques that probed the anatomical organization of white matter and gray matter in different parts of the brain.[75] Compared to TD persons, Aspies had reduced resting-state activity between part of the right TPJ and another network involved in attention, the severity of which was associated with increased emotional impairment, as measured by psychological tests. As noted, the back part of the right TPJ in neurotypical individuals is activated during social processing, and the entire right TPJ may act as a switch between the internal default and external attention network.[76] Mueller's finding is, therefore, consistent with the suggestion of Barttfeld and his colleagues that HFA/AS individuals lack balance between exteroceptive and interoceptive processing. In fact, Mueller and her colleagues went so far as to suggest that the structural and functional alterations in the right TPJ account largely for Aspies' deficits in emotional processing.[77] In light of these studies, one can't help but wonder if having a right TPJ switch that is difficult to flick might not contribute to the well-known "disconnect" between Aspies' affects and their intellectual abilities.

A small almond-shaped structure that is located within the inner surface of each temporal lobe, called the amygdala, appears to be significantly larger on the right in Aspies of various ages compared to TD persons, and to grow differently.[78] The amygdala is part of the limbic system that contributes to the processing of motivation and feelings of reward, facial expressions, fear, anxieties, and (in conjunction with nearby areas) emotional memories.[79] When viewing faces expressing fear, the amygdala of Aspies does not seem to be as integrated with nearby regions as it is in neurotypical individuals; instead it is more functionally connected with dispersed parts of the brain.[80] These wider connections likely contribute to Aspies' annoyed or fearful responses to, among other things, loud noises, bright lights, and eye gaze.[81] In light of Aspies' fondness for animals, it is fascinating that the regions of the brain that respond to faces (including the amygdala) are underactive in Aspies compared to TD individuals when processing human faces but respond similarly in both groups when viewing faces of cats or dogs.[82]

One of the most interesting neuroanatomical findings regarding AS is that Aspies have a unique pattern of connectivity within and between the two sides of their brains.[83] Children and adolescents with AS have less white matter than TD children in their right frontal lobe and right front of the corpus callosum (although the entire corpus callosum is generally found to be smaller in autistic than TD individuals),[84] whereas similar-aged individuals with HFA have less white matter in their left frontal lobe and left front of the corpus callosum.[85] Significantly, the mirrored asymmetries in the distributions of white matter in the frontal lobes and corpora callosa of people with AS and HFA are consistent with the proposal that "AS could be accounted for by right hemisphere deficits and HFA by left hemisphere deficits."[86]

The different patterns of asymmetry in the frontal lobe white matter associated with AS and HFA also fit with other neuroanatomical and cognitive differences that distinguish the two forms of autism,[87] as well as with numerous reports that the particular kinds of vulnerabilities Aspies experience (difficulty grasping metaphors, humor, figures of speech, and subtle aspects of ToM; weakened body awareness) probably entail atypical functioning in the right hemisphere.[88] For its part, the left-hemisphere deficit in white matter reported for HFA may be partly associated with the delay in the development of speech that is seen in HFA but not children with AS. Interestingly, the diminished white-matter connections across the corpus callosum associated with individuals with AS and HFA likely impedes the synthesis of information from the two sides of their brains, which may account for, among other things, their sometimes flat or inappropriate tones of voice (typically regulated by the right hemisphere) when speaking (usually facilitated by the left hemisphere).

Other findings suggest that Aspies may be impaired in multisensory integration. This is consistent not only with findings regarding diminished connectivity across the corpus callosum but also with the widespread suggestion that the brains of Aspies are relatively adept at local processing at the expense of long-distance or global connectivity.[89] A recent study that tested cognitive skills in adults with AS came to another, not necessarily inconsistent, conclusion.[90] Compared to neurotypical people, adult Aspies were slower at processing information held in working memory that is required for frontal lobe executive functions. Significantly, the authors concluded: "This particular cognitive style could argue for the reintroduction of AS in international diagnostic criteria. . . . Difficulties . . . , often masked by AS people (by their apparent fluent grammatical skills for example), may for many result in severe social isolation. As suggested by Kaland (2011), AS diagnosis will still be useful for these individuals and for providing them appropriate support."[91]

Although it is heartening that numerous researchers continue to focus on AS despite the fact that Asperger's Disorder was eliminated from *DSM-5*, new discoveries about this condition are almost always described in terms of impairments. This negative bias is unfortunate and misguided. Aspies' intellectual strengths, noted long ago by Hans Asperger, tend to favor the kinds of focused and rapid analytical processing of sequential information that is typically associated with the brain's left hemisphere.[92] (Although HFA has not been as carefully defined or studied as AS, its apparently unique pattern of brain lateralization may be associated with certain right-hemisphere strengths, such as extraordinary visual thinking.[93] More on that in the chapter 5.) Differing patterns of brain lateralization are just that—different. They may engender neurological deficits in some areas, but there may be some offsetting bonuses. In short, different is not necessarily bad.

Aspies' Neurological and Cognitive Strengths

There is no doubt that the wiring of Aspies' brains develops atypically, or that, as acknowledged above, this is linked with impairments in social interactions and, frequently, motor coordination. But focusing on the deficiencies associated with AS can be misleading when one takes an evolutionary perspective. For example, to the extent that Aspies' brains contain a greater number of tightly connected local regions at the expense of connections between areas that are further apart, they are manifesting a pattern that is at the high end of the range of variation for an advanced evolutionary trend.[94] Similarly, increased brain lateralization is an important distinction between humans and other primates, and in some ways Aspies' brains (and perhaps those of people with HFA) appear to be even more lateralized than is average for neurotypical individuals. (As the reader will recall, Aspies also tend to be at the high end of the range of variation for certain advanced evo-devo trends, discussed in chapter 1.)

But what, exactly, are Aspies' strengths? To reiterate, people with AS (and HFA) tend to have enhanced perceptual abilities, such as especially keen vision and hearing, although, as we saw in chapter 2, such hypersensitivities can be bothersome as well as a blessing. Aspies often perform exceptionally well on tests of verbal comprehension, vocabulary, and memory for facts. Some Aspies "absorb all reading material from an unusually early age" (and may, thus, be hyperlexic),[95] and research shows that adults with AS rely on different and more widespread neurological networks for reading than neurotypical adults.[96] Thanks to British psychologist Simon Baron-Cohen, Aspies are famously known for their superb systemizing skills and technical thinking styles,[97] about which we will have a lot more to say in chapter 5. Aspies are able to filter out unwanted sensory distractions by repetitive stimming or by focusing intensely on something else. They also appear to be better at thinking outside of the box than TD individuals when it comes to tests of originality,[98] perhaps partly because reduced social interaction may free them from the constraints that TD people often feel because of what others may think.[99]

But where Aspies *really* excel is in becoming intensely focused on their own special interests, as observed by Hans Asperger, many subsequent researchers, and numerous Aspies from whom we heard in chapter 2. It seems that, unlike the general population, Aspies most efficiently use their neurological networks when they are tuned to internal ruminations rather than to the external world and that, rather than activating the default network during resting states, Aspies may be "defaulting" to circuits that sustain focused inward attention on thoughts and tasks that are emotionally rewarding (such as the salience network).[100] This can lead to a mental state called "flow" that is characterized by intense concentration, lack of self-awareness, loss of the sense of time, and enjoyment.[101] The experience of flow is described by athletes as "being in the zone," by mystics as "ecstasy," and by musicians as "aesthetic rapture."[102] Activities that are conducive to flow include mindfulness, meditation, repetitive rituals, spiritual experiences, and workaholic behavior.[103]

Because flow entails being absorbed in an activity and living in the moment ("be here now"),[104] it is not surprising that it depends heavily on the brain's task-positive networks rather than the default network.[105] Interestingly, trained meditators are consciously able to increase flow (which they call "experiential focus") by flipping neurological activation away from their default systems,[106] and mindfulness-based therapy has been shown to be beneficial for high-functioning adults with ASD.[107] Further, a recent study of flow in neurotypical individuals doing mental arithmetic showed decreased activity in the default network and increased activation in parts of the brain involved with "deeper feeling of cognitive control in presence of rather high positive outcome probability,"[108] which is not a bad approximation of a likely frequent state of Aspies.

Experiencing flow is probably a uniquely evolved human adaptive trait (at least among primates) that occurs to varying degrees in different people. Until recently, at least, individuals who invented major breakthroughs were likely to have had especially original, creative, and technical minds that facilitated intense focus on task-positive goals. Although the extent to which "garage inventors" are responsible for current breakthrough inventions is a subject of debate,[109] even investigators who believe that lone inventors have become eclipsed by superior innovations of teams and organizations acknowledge that "each generative insight occurs within a single mind."[110] Like their modern counterparts, the prehistoric individuals who were responsible for intellectual breakthroughs may have had a tendency to become enthralled with their particular interests.

Unfortunately, although scientists are making progress on deciphering the neurological substrates of Aspies' atypical social processing, there does not yet seem to be much exploration of the neurological bases that underpin their intellectual strengths, or those of people with HFA for that matter. (The relevant findings described in this chapter were painstakingly ferreted out from the literature.) This is a shame, not only because the advanced technical, creative, and linguistic skills that frequently occur in people with AS are prime examples of higher cognitive abilities that emerged during human evolution but also because it is high time that Aspies and individuals with HFA were given their due.

What Aspies think about when their minds are left to themselves is a tantalizing question that, as far as I know, has been pursued with only a few adult males.[111] Below, Eve addresses this question. From her narrative, it is abundantly clear that she spends huge amounts of time turned inward. By the time she reached elementary school, she was already telling herself an ongoing story about an imaginary world in which she was in charge. The story that Eve is currently telling herself has been progressing for at least a decade and, most of the time, she seems quite happy to live in this elaborate imaginary world. As you will see, Eve's thinking, in general, is quite verbal and has visual and emotional elements. When her mind wanders, it seems to focus on practical or mundane matters with very little interest in past or future social interactions in the external world.

Unlike my sisters and other typically developed people, I didn't automatically soak up what I needed to interact with others. I was too distracted by the sounds and the things that my relatives kept in their houses to actually watch them and figure out why they acted the way they did, as most children seem to do. The only reason I am now able to interact with people in a polite manner is because I have parents and teachers who gave me rules for behavior. I can talk to people, I can even interact with them up to a point, but I will never understand their emotions because I critically lack what most people call empathy. I usually have to ask why someone is upset if I see them crying or just showing some other overt sign of emotion. I cannot figure out for myself what has happened around that person to elicit that emotional response. I can't imagine their circumstances either.

For me, words are a powerful thing, and they help keep my world in order. An inner voice ever narrates or directs my actions. Whose voice speaks depends on the thoughts being verbalised.[112] In the case of walking directions to a place I'm going, it is usually the voice of the person who originally showed me where the place was, by which I mean it is usually my father, Aidan's, voice. Sometimes it is my own voice if I have found a new route to or from the place I'm going. In the case of doing tae kwon do patterns, learning a new kick, or trying to answer questions about the meanings of different moves, it is usually my *sabum*'s [teacher's] voice. In some cases this inner speech verbalises the advice or directives of my many relatives when I find myself repeating a past mistake. There are some things I've been told so many times the words of the warnings and pieces of advice might as well be printed in bold on my head in permanent ink. The specific words would be something along the lines of "No smoking, drink only a little, and no drugs whatsoever" as well as "Slouching is bad for your spine; look me in the eye; and don't speak while you are eating, it makes you look like a chipmunk."

When I let my mind wander, I usually think about the weather, the book I am reading, how I feel physically, or fanfiction stories that I tell myself based on different anime or manga I've seen. I also occasionally get snatches of songs. Here is a sample of the thoughts that would probably rampage through my head whilst I walk to my tae kwon do class in the rain:

AAAAAAHHHHHH, it's raining. I don't have an umbrella. So cold. My book is soaked. I hope it doesn't fall apart. EEEEEYUCK, I'm soaked. Why does this *dobok* [martial arts uniform] have to be made of cotton and polyester? It's seriously heavy when wet. Agh, my shoes and feet are soaked. WHY does it have to start raining like this? Damn English seesaw weather, it was so sunny earlier. (Beginning lyrics of "Ironic" segueing into "Swing Low, Sweet Chariot" or "Lean on Me," depending on my mood.) This humidity is really getting to me. I hope he doesn't have us sparring tonight. Wet armour isn't going to be comfortable. Not that this stuff is comfy even when dry. Cross the road and walk up the hill on the left side. Cross again at the top of the hill and go left until you come to the turning that goes down the hill after the derelict church. Agh, I hope this rain stops soon.

Not all my thinking is verbal. As I mentioned earlier, I tell stories to myself out loud, but in a volume that other people aren't meant to hear. I visualise the settings and characters of these stories in 3D animated colour. For example, in the story that I have been telling myself for at least ten years, which I call "Firefox Chronicles," my main character lives in a country that is similar to a hotter version of Japan with the buildings coloured red, and his hair and eyes change colour depending on what personality he is manifesting. He wears traditional Japanese robes in a somewhat girly style due to the fact that in the main world that my story is set in, women are the rulers and men rarely ever get to take control even at a minor level. Much to the point that my main character, who is a boy, dresses as a girl in order for people to take him seriously as the heir to the throne and takes lessons in both boyish and girlish modes of behaviour. He can also use magic that is said to be used only by pure girls and never by boys. My main character can summon the Goddess of Light from the Divine World, the God of Death from the Lower World, and can use their powers as his own because they chose him as their vessel on his thirteenth birthday.

The story helps me relax and block out some of the more abrasive sensations of the real world, such as brightness and coldness. I never have any idea where the plot is really going because I haven't planned out the story or stories to that extent. I usually just have my main characters and their companions either hopping from world to world in different animes and cartoons I've seen, playing a game in the home world I semicreated for them, or navigating some sort of danger in that world. I also introduce new characters of my own creation into the worlds that I have my main characters going to.

I'm not sure how long I will continue telling myself the "Firefox Chronicles" because there is no clear beginning or end to the tale, and the plot is wandering lost down a country road. I really see no reason to rescue it. After all, it's my story and I'm only telling it for the fun of it to an audience of one that really doesn't care whether the story is coherent, cohesive, or convincing in the least. It's something to do in my spare time when I feel the world is becoming too close for comfort and needs to be pushed back several million paces. If I ever decide to write down my favorite fanfiction and make it official, that's when I'll make the plot somewhat cohesive and coherent. Until such a time occurs, the "Firefox Chronicles" are merely mine to manipulate as I please, and no one can really tell me which part I may put where.

As to emotional thinking, emotions aren't something that you have to think about, they are something you feel or experience. Thinking about certain things can evoke a physically emotional response for me, though most of those thoughts are usually stressful or nervous, such as in the case of waiting for my computer to be repaired. Other things that evoke emotional responses are books. Because emotions and words are connected, there's no way to detach them from each other. To be more specific, paranormal romance, fantasy, and young adult books evoke big emotional responses from me, such as fear or breathless anticipation, though I also feel some measure of relaxed comfort if I am reading a book I've read before or the continuation in a series that I really enjoy. Books have a negative effect on my emotions if I can't concentrate on the plot. Although I love classics like *The Odyssey* and *The Iliad*, classics that don't involve fantasy on some level tend to bore the bones out of me.

To summarise, I am not generally all that social with other people. The majority of the time I am using verbal thinking with inner speech as my medium to tell me where I'm going, what I'm supposed to do, and the expected behaviours for when I get there, which could involve respecting my instructor, or remembering homework, what room a particular lesson is in, and other bits and pieces of information. As for thinking in images, I'd say that the time I use that is usually when I'm telling the "Firefox Chronicles" to myself, which makes the story so vivid and real that I sometimes wish I lived in the world that I created.

Eve at a few days
old, with her father
(Aidan) and mother
(Sarah), July 1991.
Notice the book!

Sarah with five-month-old Eve,
December 1991.

Eve at five months. As one friend remarked, "A striped baby on a striped couch."

Eve at age three and a half "reading" to Grandma Dean and sister Helen.

Helen on the left; Eve on the right. Eve has been fond of animals since she was a little girl.

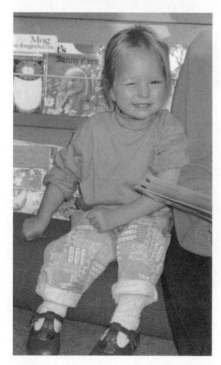

Eve at age three and a half in front of the books at her Montessori school. Notice the squint, which may have been related to her development of brain lateralization.

Eve at age three and a half. She pointed a lot, which is thought to be related to development of a "theory of mind."

Eve at age four.

Eve, seven, and Helen, five, ready for dance class. Although Eve wasn't very coordinated, she liked to go to class with Helen.

Eve at age seven and a half, ready to go to the Nutcracker ballet with Judith, Helen, and Grandma. Eve found the Christmas dress scratchy but was determined to wear it.

Eve's sisters, Helen and Judith, have always looked out for her.

Eight-year-old Eve in a cozy space.

Eve, Helen, and Judith at their mother's (Sarah's) 1999 graduation from law school.

Eve says she "had a good and fun childhood." Here she is at a "tea party" with Helen and Judith, 2002.

Eve, age twelve, with father (Aidan) and sister (Judith), 2003. Notice the book.

Twelve-year-old
Eve with her doll
collection.

Eve, age nineteen, with a horse in Santa Fe, 2010.

Eve doing her thing, 2010.

Eve at age twenty-one.

Nice doggy. Eve visits a
friend's dog, 2012.

Eve has loved spicy food ever since she was a little girl. Here she is at a restaurant, 2013.

Eve and Grandma Dean, 2013.

Eve graduates from Bath Spa University in July 2016.

Eve at her gradua-
tion, 2016.

Chapter 4

Sex and Gender in Asperger Syndrome

The autistic personality is an extreme variant of male intelligence.

—HANS ASPERGER[1]

According to autism expert Tony Attwood, the "overwhelming majority of referrals for a diagnostic assessment for Asperger's Syndrome [AS] are boys."[2] Attwood reports the ratio of male to female Aspies to be about 10:1, although some researchers believe that the ratio may be appreciably lower.[3] Nevertheless, Attwood's estimate is consistent with reports that the percentage of males increases at the highest functioning end of the autistic spectrum and among individuals with the highest IQs.[4] Interestingly, not only are there fewer females diagnosed with AS, but they also tend to be identified later than boys, at around age nine versus age seven.[5]

Experts reason that the relatively low proportion of females diagnosed with autism, especially AS/HFA, is due partly to diagnostic criteria that have been biased toward males since the time of Hans Asperger himself.[6] For example, a study of thirty-three adult males and twenty-nine adult females with AS or HFA, who had equally severe core symptoms as children, found that adult females reported more lifetime unusual sensory experiences and self-identified autistic traits than the males but significantly fewer difficulties in interpersonal interactions and less autistic behavior when it came to repetitive stereotypical behaviors. The two groups did not differ in other ways, however, such as self-reported empathizing and systemizing traits.[7] The authors noted that the contrast they documented between females' childhood symptoms and their adult interpersonal skills was consistent with numerous reports[8] that, compared to their male counterparts, females with ASC (autism spectrum conditions) may be more motivated and may put more effort into developing compensatory skills that help them to appear 'socially typical.' . . . Indeed, experienced clinicians have observed that one reason females (girls or women) with ASC may be less easily identified is because of their ability to 'camouflage' their autism. This type of camouflaging may involve conscious, observational learning of how to act in a social setting and by adopting social roles and following social scripts. Hence, a female teenager or adult with ASC may be able to develop reciprocal conversation, social use of affect, gestures and eye gaze, that would place them under the radar for the more commonly understood and recognizable (male) phenotype of ASC. Some of the women with ASC reported they consciously 'cloned' themselves on a popular girl in their class whilst at school. . . . Women who adopt these camouflaging strategies nevertheless report that underneath their superficially sociable behavior they are often experiencing high levels of stress and anxiety as they have to work hard to keep up the mask.[9]

Because these findings concern only the small percentage of females who have actually been diagnosed compared to males, it is easy to see why experts believe there may be more AS/HFA females out there (perhaps with less severe symptoms) who have not been diagnosed because of inadequate diagnostic criteria combined with what has been dubbed the "female camouflage effect."[10] It also makes perfect sense that researchers and clinicians are calling for "development/revision of diagnostic assessment instruments to address the sex/gender differences in ASD manifestation."[11]

As important as they are, diagnostic criteria and female camouflaging account only partly for the skewed sex ratio for AS, which appears to stem from a number of overlapping causes, including biological factors that might selectively protect females (such as females having protective genes on at least one of their two X chromosomes).[12] As detailed below, the brains of male and female Aspies are also wired somewhat differently, which is likely related to how fetal exposure to prenatal hormones organizes development in the brains of the two sexes. Despite the fact that Aspie boys and girls share certain challenges and cognitive abilities that set them apart from neurotypical individuals, the intense interests developed by males and females with AS tend to differ in predictable and interesting ways. However, a discussion of both the similarities and differences between male and female Aspies will make more sense if we first review the contentious topic of sex/gender differences among typically developed individuals.[13]

Nature, Nurture, and the Battle over Sex Differences

In addition to well-known differences in physical attributes such as average height, weight, and voice pitch, TD males and females around the world differ in an estimated sixty-five behavioral and cognitive traits, although most of these disparities are minor in nature.[14] Some of the more obvious differences involve how the sexes experience fear, anxiety, and depression; express violence and aggression; and react to risk. Neurotypical males tend to have stronger systemizing skills, whereas females are generally better at empathizing, as famously documented by British psychologist Simon Baron-Cohen.[15] It is also widely documented that males excel, on average, at visuospatial skills such as mental rotation,[16] whereas females tend to be better at grasping and producing linguistic (including reading),[17] gestural, and emotional communications. Consistent with this, "men and women show very large differences in their gender-related interests, with men more inclined toward thing-oriented activities and occupations (e.g., mechanics, carpentry, engineering) and women more interested in people-oriented activities and occupations (e.g., counseling, elementary school teaching, nursing)."[18] Further, these sex differences are "significant and consistent across 53 nations. . . . About the same magnitude as sex differences in height."[19]

Regrettably, academics have become entangled in a polarizing debate about the causes of the differences between males and females.[20] At one extreme, many social scientists and feminist scholars are unreceptive to the possibility that biology or evolutionary history might contribute to the average behavioral and cognitive differences between the sexes. Instead, they view the differences as solely the result of sex-role socialization and male sociopolitical dominance (nurture). At the other extreme, some evolutionary biologists tend to interpret sex differences as strictly the result of evolved adaptations (nature), dismissing nurture arguments as scientifically uninformed "political correctness." There is nothing in evolutionary theory, however, that precludes sex-role socialization and male-biased political power from contributing to sex differences, or that justifies sexist behavior or prevents it from being redressed. For its part, "feminism will have to for-sake this hostility [toward evolutionary theory] if it wants to retain its intellectual credibility."[21] Fortunately, feminist and evolutionary perspectives are beginning to be reconciled, thanks to reflective scholars who incorporate elements from both schools of thought.[22]

The size of sex differences varies for different features, both within and across societies, depending on cultural, social, and environmental conditions, as emphasized by nurture theorists and confirmed by evolutionary biologists. Nonetheless, persuasive cross-cultural evidence suggests that, whatever their immediate ("proximate") causes, many, if not most, universally identified sex differences are rooted ultimately in evolutionary adaptations.[23] Indeed, sex differences are pervasive among mammals for the very reason that they maximize the chances for individuals to reproduce successfully, which is what evolution is all about.[24] For males, this means locating and inseminating partners, which is facilitated in many species by, among other qualities, good navigational skills, strength, and assertiveness. For females, including those among higher primates like ourselves,[25] it means conceiving, bearing, and raising healthy offspring, which requires mating with suitable (fit) males, acquiring adequate nutrition to support gestation and nursing, and being attentive mothers once infants are born. As evolution proceeds, physical attributes and behaviors in each sex that increase the number of offspring some individuals produce relative to others are passed on to their descendants and, thus, spread and are transmitted to future generations (that is, they are naturally selected).

Scientists refer to species in which males and females noticeably differ as sexually dimorphic, which literally means that the sexes manifest two (di) average forms for certain traits, although the ranges of variation overlap in males and females for most, if not all, these traits. For example, humans are among the moderately dimorphic primates when it comes to stature and weight. Although men are taller and heavier on average than women, stature and weight nonetheless overlap extensively in the two sexes, which means there are plenty of men and women who match the heights and weights of members of the opposite sex.

Despite the fact that scientists often report average sex differences for single features such as body size or hormone levels, sexual dimorphism encompasses many variables that hang together because they were (and may still be) adaptive from an evolutionary perspective. In humans, for example, a higher level of testosterone in male fetuses sets the stage for the development of (among other dimorphic features) larger, taller, and more muscular adult bodies that are often attractive to, and probably aid in the competition for, women.[26] A testosterone-mediated deeper male voice (which develops during puberty) is also part of this package. Lower voices tend to be attractive to women in addition to being correlated with larger male bodies (particularly upper musculature), which makes them likely vocal signals for intimidating potential rivals, as is the case for many mammals.[27] The evolutionary importance of deep male voices is illustrated by the Hadza, a human population of hunter-gatherers who live in Tanzania, among whom men with the lowest voices have been documented to have more children.[28]

But it wasn't just big bodies and deep voices that were selected for in our male ancestors. Enhanced visual cognition and systemizing skills in prehistoric men likely facilitated navigational skills and hunting-related abilities, including "tracking, aiming, throwing, geometric analysis of spatial relationships, and environmental sound analysis,"[29] which helped them locate and attract mates as well as provide for offspring. For their part, rather than growing big bodies, prehistoric women used their calories to gestate and nurse offspring, as women do today.[30] Additionally, sharpened communication and empathizing skills in hominin (and contemporary) women would have facilitated reciprocal interaction with their infants, hence contributing to successful mothering and therefore infant survival (not to mention other advantages discussed in chapter 1, such as language evolution). In sum, natural selection for dimorphic body builds and cognitive skills, on average, likely increased the number of viable offspring for both prehistoric males and females, which, as noted, is the essence of evolution.

Differences in the Brains of TD Males and Females

The dimorphic behavioral and cognitive attributes that typify humans are underpinned by fascinating neuroanatomical and neurochemical differences in the brains of males and females, despite a common misconception to the contrary.[31] Men have larger brains than women, both absolutely and relative to a given body size, although there is extensive overlap in brain size between the two sexes. This is not simply because males are taller and have bigger bodies than females; men have brains that are roughly 10 percent larger than those of women whose bodies are the same size.[32] One, therefore, should not simply compare the absolute sizes of the brain or its parts in the two sexes without adjusting for the fact that males have bigger brains, on average, to begin with. One way to do this is to compare brains that are the same size in men and women. When this is done, important sex differences that are independent of brain size are revealed. For example, although same-sized brains of men and women do not differ in the total amounts of their gray and white matter,[33] contrary to what many believe,[34] women's brains average greater amounts of gray matter in certain regions,[35] whereas men's brains have relatively more white matter in some places.[36]

Although a number of other structural differences have been documented for the brains of neurotypical men and women,[37] we need to keep in mind that these differences are averages, and that anatomical features in the brains of the two sexes overlap extensively—so much so, in fact, that it would be a mistake to view human brains as sorting into two categories—male brains and females brains.[38] The same may be said when it comes to the many sex differences, on average, in behavioral and cognitive attributes. Although our species is definitely sexually dimorphic, men and women, again, do not form two distinct groups (despite the familiar quip that "men are from Mars, women are from Venus").[39]

Despite these caveats, the wiring of men's and women's brains differs, on average, in striking ways that were likely inherited from our early hominin ancestors. For example, one study of resting-state functional connectivity in neurotypical individuals revealed significantly different patterns of brain lateralization for men and women, which may reflect a sex difference in specialization and integration within and between the two hemispheres.[40] This suggestion is bolstered by the results of a recent analysis of white-matter connections, which found "conspicuous and significant sex differences that suggest fundamentally different connectivity patterns in males and females."[41] Connectivity in the brains of males proved to be optimal for communication along the anterior-posterior length within each hemisphere, whereas for females, the connectivity was relatively enhanced between the two hemispheres, that is, across the brain rather than from front to back within each side (fig. 4.1).

The brains of males also have more short-range subnetworks within lobes

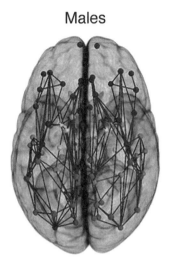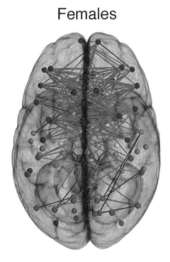

Figure 4.1 Conspicuous brain connections that distinguish the brains of neurotypical males and females. Males have greater front-to-back connectivity within each hemisphere of the cerebral cortex compared to females, whereas females show relatively more connectivity across the midline between the two hemispheres. This suggests that female brains are relatively adept at communicating between regions in the left and right hemisphere that process analytical and intuitive information, respectively. Brains are viewed from the bottom with their right sides on the reader's left. Based on data for 428 males and 521 females, eight to twenty-two years old. Reproduced from Ingalhalikar et al., "Sex Differences," fig. 2A.

compared to female brains, which have relatively enhanced long-range connections.[42] These remarkable differences led scientists to conclude that the connectivity patterns in male and female brains might "confer an efficient system for coordinated action in males. Greater interhemispheric connectivity in females would facilitate integration of the analytical and sequential reasoning modes of the left hemisphere with the spatial, intuitive processing of information of the right hemisphere."[43] Interestingly, these findings are consistent with the fascinating discovery that men and women use different networks in their brains when engaging in mental rotation of geometric figures.[44]

The richer structural connectivity between the two hemispheres in females is consistent with reports that women have somewhat larger corpora callosa than men after adjusting for brain size, in addition to leftward asymmetry of frontal lobe white matter (all the better for linguistic processing).[45] It is also noteworthy that functional connectivity is especially dense within certain parts of the right hemispheres of males, which probably contributes to their comparatively better-than-average visuospatial skills as well as greater vulnerability to disorders that disrupt typical patterns of brain lateralization.[46]

The sex differences that develop in the brains and behaviors of people do not emerge on blank slates, of course.[47] Instead, genetic factors (such as having or not having the male Y chromosome) together with exposure to variable amounts of prenatal hormones contribute significantly to these differences. For example (and of particular relevance for this chapter), exposure of male fetuses to relatively high doses of testosterone compared to female fetuses appears to "masculinize" their brains in ways that promote the eventual development of male-typical behaviors such as risk taking, sensation seeking, and a tendency toward systematic rather than empathetic thinking.[48] Female fetuses are also exposed to fetal testosterone, but usually to a lesser degree than developing males. However, in cases within the general population, where either boys or girls are exposed to relatively high doses of testosterone (measured from amniotic fluid), both grow up to engage in increased amounts of male-typical play compared to their same-sex peers.[49] This is an important finding because it is the "first documentation that androgen exposure prenatally relates to sexually differentiated play behavior in boys and in girls."[50]

Sex Differences in Autism

Do autistic individuals manifest the sex differences that appear to be universal among neurotypicals, or does having autism somehow remove, attenuate, or alter these differences? The scientist who has most prominently addressed this question is Simon Baron-Cohen. For starters, he has shown that males and females with autism, including those with AS, share an atypically high ability for systemizing, defined as the

> drive to analyze or construct a system—a mechanical system (such as a car or computer), a natural system (nutrition) or an abstract system

(mathematics). Systemizing is not restricted to technology, engineering and math. Some systems are even social, such as a business, and some involve artistic pursuits, such as classical dance or piano. All systems follow rules. When you systemize, you identify the rules that govern the system so you can predict how that system works. This fundamental drive to systemize might explain why people with autism love repetition and resist unexpected changes.[51]

Baron-Cohen and his colleagues showed that males and females with autism score lower on average than neurotypical people when it comes to comprehending and feeling another person's thoughts and emotions (as measured with the empathy quotient, or EQ).[52] Because neurotypical males score significantly higher, on average, on tests of systemizing (SQ, or systemizing quotient) but have significantly lower average EQs than neurotypical females, Baron-Cohen has suggested that the high SQs and low EQs of males and females with AS and high-functioning autism indicate they have "extreme male brains."[53] Baron-Cohen even went so far as to wager a drink that "females with autism [will] show a male pattern of brain activity when we use functional magnetic resonance imaging" to observe their brains.[54] (More on this below.)

A recent extension of the extreme male brain theory holds that females with autism are "masculinized" at more pronounced levels than males with autism.[55] This is akin to suggesting that the degree of sexual dimorphism may be altered in autism compared to the general population, a provocative idea that deserves close scrutiny. Although it is generally agreed that sexual dimorphism is due largely, but not entirely, to the effects of prenatal and (later) pubertal hormones, especially testosterone,[56] one should investigate multiple rather than single traits that are theoretically dimorphic because of hormonal influences in order to assess the suggestion that female Aspies are pervasively masculinized, something few autism researchers do.

A notable exception is Susanne Bejerot of the Karolinska Institute in Stockholm, Sweden. She and her colleagues compared male and female adults with AS or HFA to TD adults with respect to hormone levels, physical traits, psychiatric profiles, sexuality, and gender roles—and incorporated an evolutionary perspective to boot.[57] The authors found that AS and HFA women do, indeed, have elevated testosterone levels as well as a few more traits typically associated with masculinity than TD women, but they also discovered that AS and HFA males have more typically feminine character- istics than TD men, leading the authors to conclude that Autism Spectrum Disorder is a "gender defiant disorder" that is associated with androgynous features in both sexes rather than excessive masculinization of the brain.[58] As the next section shows, Aspies manifest some interesting forms of sexual dimorphism, as compared to neurotypical people.

Sexual Dimorphism in Aspies

The most obvious sexual dimorphism that characterizes higher primates, including people, is that males are, on average, bigger-bodied than females. If Aspie women are generally masculinized, as some believe, one would pre- dict that their weight relative to height (body mass index) would be greater than that of neurotypical women. Similarly, if Aspie men were hypermascu- linized, one would expect their relative body weight to be larger than that of TD men. Neither prediction seems to hold. Thus the body mass indices of Aspie women did not differ significantly from those of TD women,[59] whereas those of Aspie boys and adolescents were significantly lower than those of TD males.[60] If anything, available information about relative body size in Aspies suggests that, rather than females being relatively masculinized, males are relatively feminized. Sexual dimorphism in body weight may thus be reduced in AS compared to TD individuals.

The findings are not as clear-cut when it comes to blood serum chemistry, however, as shown by a study that analyzed serum components in male and female adult Aspies and compared them with those from neurotypical men and women.[61] Aspies of each sex had numerous serum biomarkers that were distinct and overlapped minimally between the two sexes. These sex-specific biomarker fingerprints also differed from those of same-sexed TD individu- als. In other words, Aspies' blood serum chemistry is sexually dimorphic and unique compared to the general population. Nonetheless, Aspie women had elevated levels of biologically active testosterone, consistent with the extreme male brain theory.

The blood serum study incorporated measurements of each participant's number of autistic traits (or AQ, autistic quotient), as well as measures of their adeptness at empathizing (EQ) and systemizing (SQ). Curiously, this research found that Aspie females averaged the highest AQ and SQ scores among all the participants—even the Aspie males, which reversed the direction of the sexual dimorphism in these traits compared to TD individuals. Aspie females also had the lowest average EQ scores, including compared to those of Aspie men, which, again, reversed the expected direction of this sex difference. The extremely low average EQ and extremely high average AQ and SQ scores of Aspie women were interpreted as evidence that they have "more 'masculinized' scores" than both TD females and males with AS.[62] The other side of this coin, of course, is that AS men appear to be relatively feminized compared to AS women.[63]

What about Baron-Cohen's wager that autistic females would eventually be shown to have more masculinized brains? Does this prediction hold for Aspie women? One investigation of total gray and white matter found that the brains of Aspie men and women did not differ in their total amounts of gray matter, contrary to the general population in which male brains have absolutely more gray matter.[64] In terms of total white matter, however, the Aspies showed the expected dimorphism, with men having more than women, but the difference between the two sexes was not as large as in TD individuals. Although the total size of the brain was not taken into account, the authors made a convincing case that the sexual dimorphism found in total gray and total white matter of neurotypical individuals was absent or attenuated in the adults with AS they studied. They further concluded that "no evidence of hyper-masculinization was found, indicating the previously described behavioral and neuroendocrine effects do not map onto anatomic features in a straightforward way."[65] (Baron-Cohen, incidentally, was a collaborator on this study.)

One must be cautious, however, about applying the findings regarding sexual dimorphism in the total amounts of gray and white matter in Aspies to individual components of their brains. For example, limited evidence suggests that, when it comes to the white matter connecting the two hemispheres through the corpus callosum, the brains of Aspie women may resemble those of TD women in being denser and more efficient compared to the brains of their male counterparts.[66] This possible similar sexual dimorphism in connectivity through the corpus callosum of Aspie and TD individuals is especially interesting in light of figure 4.1, although it is unfortunate that there is not yet enough information to compare the extent of the dimorphism in the two groups.

Do Aspie males and females have different ways of thinking, on average, during certain mental activities, similar to neurotypical individuals? One of the most dramatic cognitive sexual dimorphisms in TD individuals is the increased average ability of men to rotate visual images mentally compared to women (see fig. 4.2), a skill that appears to be associated with enhanced systemizing abilities and intuitive physics.[67] Although TD men and women both show increased activity in their parietal lobes during mental rotation, especially on the right side as one would predict, women in addition use part of the frontal lobes on both sides.[68] Intriguingly, the part of the left frontal lobe that is activated during mental rotation in TD women overlaps with the speech areas in that hemisphere.[69] The fascinating possibility has therefore been suggested that, in addition to processing visual gestalts during mental rotation, women may speak silently to themselves more often than men when trying to rotate images.[70] (I know I do!)

What about mental rotation in Aspies? Are males better at it, on average, than females? Unfortunately, although numerous studies have compared visuospatial skills between autistic and TD individuals,[71] these investigations failed to compare the mental rotation skills of male and female autistic individuals, including Aspies. What they discovered, however, is that autistic people are somewhat better than neurotypical individuals at discovering figures that are hidden within images and recreating geometric designs using blocks, probably because autistic individuals tend to be especially good at processing small parts of visual images compared to the overall big picture.[72] Interestingly, TD males are somewhat better than TD females at finding hidden figures, but there is no known neurotypical sex difference in performance on block designs.[73] With respect to mental rotation, Aspies sometimes outperform TD individuals, but when they do it is because of enhanced skill at the nonrotational aspects of the tests (e.g., comparing one stationary figure with others) rather than the actual mental twirling of images.[74] These studies converge on the conclusion that, although individuals with AS and HFA have somewhat different skill sets than TD folks when it comes to visuospatial processing, the package of visual skills associated with autism is not what one would predict from the extreme male brain theory.[75]

Although it is unclear whether Aspies share the TD proclivity for sexual dimorphism in mental rotation ability, an important study of brain activation in Aspie men and women found significant sex differences across widespread

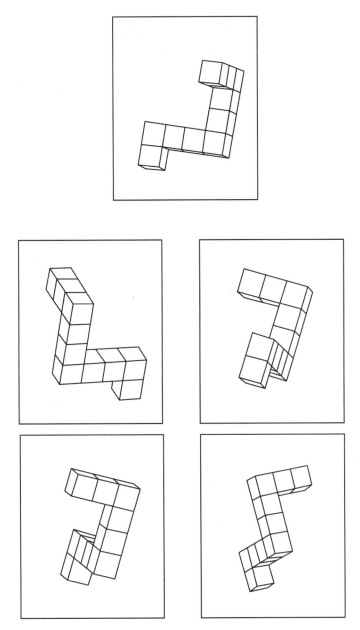

Figure 4.2. Mental rotation task. Judge which two of the four figures on the bottom are rotated versions of the one on the top. This is a relatively easy example of the task. Males, on average, outperform females on mental rotation exercises, especially on more difficult ones. (Answer: 1 and 3.) Courtesy of Michael Peters (Peters et al., "Vandenberg and Kuse Mental"; Peters and Battista, "Mental Rotation Stimulus Library").

parts of the brain during relatively easy mental rotation tests on which the two sexes performed equally well.[76] This study also compared the patterns of sexual dimorphism found during mental rotation in Aspies and TD individuals, with fascinating results. In both autistic and TD individuals, the same specific regions of the brain became activated during mental rotation, and there was sexual dimorphism in the relative strength of the activity in each region. However, in each area the direction of the sexual dimorphism was reversed in the two groups. As just one example, a part of the frontal lobe that is more active during mental rotation in TD females compared to males had the opposite pattern in Aspies (that is, its activation was stronger in AS males than females). The bottom line is that there are significant sex differences among both AS and TD individuals when it comes to processing mental rotations, but their patterns of sexual dimorphism are very different.[77]

What all these studies boil down to is that, although Aspies appear to be sexually dimorphic in body weight, blood chemistry, aspects of brain connectivity, and performance on certain cognitive tasks, these dimorphisms are not simple extensions of those found in TD folks, nor do they support the idea that Aspies are generally hypermasculinized. As we have seen, the direction and extent of the sex differences in many features associated with AS is complex, resulting in an overall pattern that is completely different from that of neurotypicals. As the next section shows, however, gender-related interests among people with AS are broadly consistent with neurotypical females' superior linguistic and reading skills (chapter 1)[78] and the worldwide trend for males to be inclined toward thing-oriented activities whereas females gravitate toward more nurturing and humanistic interests.[79]

Sugar and Spice and Everything Nice— That's What Aspie Girls Are Made Of

Despite their mutual strength in systemizing abilities, the interests and occupations chosen by males and females with AS tend to be different, as detailed by Tony Attwood:

> The most popular special interests of boys with Asperger's Syndrome are types of transport, specialist areas of science and electronics, particularly computers. . . . Girls with Asperger's Syndrome can be interested in

the same topics but clinical experience suggests their special interest can be animals and classic literature. These interests are not typically associated with boys with Asperger's Syndrome. The interest in animals can be focused on horses or native animals and this characteristic dismissed as simply typical of young girls. However, the intensity and qualitative aspects of the interest are unusual.[80]

Elsewhere, Attwood notes that Aspie girls frequently substitute animals for human friends, probably because animals "do not deceive, tease or behave in as fickle a way as occurs with humans, and are non-judgemental."[81]

Aspie girls also create substitutes for friends in other ways. For example, a girl with AS may have a special interest that is typical of the interests of girls in general,

such as collecting Barbie dolls, but the girl . . . may have a collection of many more dolls than her peers, the dolls are arranged in a particular order and she usually does not share her play Barbies with a friend. . . . The dolls can also be used to rehearse what to say in prospective situations, and can become alternative friends who, perhaps unlike real girls in her life, are supportive, inclusive and kind. The interest is solitary and functional.[82]

Girls are also more likely to enroll in speech and drama lessons, where "knowing a script . . . means the child does not have to worry about what to say."[83]

One of the most poignant examples of how girls with AS cope with social isolation is by escaping into alternative worlds through literature. According to Attwood, "Teenage girls with Asperger's Syndrome can develop a fascination with classic literature such as the plays of Shakespeare and poetry. Both have an intrinsic rhythm that they find entrancing and some develop their writing skills and fascination with words to become a successful author, poet or academic in English literature."[84] Aspie females often inhabit alternative worlds through science fiction and fantasy, which may include fairies, witches, and mythical monsters. Remarkably, these alternative worlds sometimes include specific cultures or periods of history, such as Scandinavian Vikings, ancient Egypt, or "Japan (a country renowned for its fascination with technology, and the creation of cartoon characters)."[85]

When I first read Attwood's observations about Aspie females, I was surprised at how specific they were, given the relative rarity of females with AS. What amazed me even more was that Eve had *all* the attributes that Attwood described for Aspie females, as you have seen or will see from her parts of this book. She collected storybook dolls as a child, which she kept neatly in their boxes (and still does), and she participated in drama groups for the reasons Attwood mentions. She also loves all things Japanese (and collects manga [Japanese comic book stories]) and prefers the company of animals—again, for reasons that are consistent with Attwood's analysis. Most of all, Eve is an avid reader with a love of fantasy literature that is all-consuming. (Recall also from chapter 1 that Eve has a "slight belief in fairies" because she was told as a child that she was a changeling.) As she has observed, reading provides alternative worlds that are more comfortable for her than the real one.

But Eve's not the only Aspie who fits Attwood's female profile to a tee. Nadia (a pseudonym) is an eighteen-year-old woman from Australia who reports that when she was younger, she coped with the stress of being at school by reading, which minimized the need to interact socially with peers.[86] She had precocious language development, is highly verbal, and enjoys acting in drama productions. Nadia escapes into fantasy literature, and, like Eve, comments that "once I've read a book I want to be able to imagine situations that didn't happen. I want to imagine a life for these characters that never happened in the book."[87]

Another Australian woman, forty-eight-year-old Kahla (a pseudonym), was also precocious in her development of language and reading. Kahla is intensely interested in other cultures and (get this), as a child "entertained the idea that perhaps I was a changeling, that I'd been swapped at birth and the real me was being looked after by fairies and I was really some kind of strange fairy child that had been left for my parents to raise."[88]

These women also have two other interesting traits in common that are not included in Attwood's profile: they are left-handed and all three habitually talk aloud to themselves. The women's left-handedness may not be too surprising, given that the occurrence of this trait is often reported to be markedly higher in autistic than TD individuals.[89] Thus whereas left-handedness is typically found in about 8–10 percent of TD individuals, it seems to occur in twice as many individuals with ASD (and sometimes many more).[90]

The fact that Nadia, Kahla, and Eve all speak aloud to themselves is extremely interesting and raises the question of whether this is typical of other Aspie females. Recall from chapter 1 that Eve talks to herself "so that my logic can keep up with my senses or something like that, or maybe it is just that the talking to myself helps me to make sense of the stimuli in the world around me." Kahla and Nadia similarly explain that hearing themselves speak aloud lets them know what they think.

Curiously, the outspokenness of all three women is reminiscent of individuals who have undergone "split-brain" surgery (a corpus callosotomy), which severed their corpora callosa down the middle as a last resort to treat the most extreme forms of epilepsy. Early research on split-brain patients, pioneered largely by neuropsychologist Roger Sperry, confirmed that the two halves of the brain function differently (fig. 3.3) and, further, demonstrated that individuals with severed connections between the two hemispheres appear to be of two distinct minds rather than one. (Sperry shared the 1981 Nobel Prize in Medicine or Physiology for this research.) The experimental methods used to discover the details of brain lateralization depended on accessing one side of patients' brains to the exclusion of the other, which was tricky partly because, without realizing it, patients were very good at cluing in both sides of their brain to information that was initially received by only one hemisphere. One way they did this was by speaking out loud so that the information would go into both ears and thus reach both sides of the brain. As Sperry put it, "In many tests the major [language] hemisphere must be prevented from talking to the minor hemisphere and thus giving away the answer through auditory channels."[91] Could it be that Aspies who talk aloud to themselves are responding to relativity weak connectivity between the two sides of their brains and/ or possible atypical neurological lateralization for language?[92]

It should be clear from these last two chapters that AS is associated with uncommon patterns of brain connectivity and brain lateralization, and that these differ to some extent between males and females. As we have seen, connectivity and lateralization in the brains of neurotypical individuals is also sexually dimorphic, albeit in ways that often differ from the dimorphism seen in AS. A possible exception is that, although the entire corpus callosum is generally found to be smaller in autistic than TD individuals,[93] its connections between the two sides of the brain may be relatively enriched to varying degrees in both Aspie and TD women compared to their male counterparts.

If so, one wonders if this structural feature promotes the development of greater functional integration between the two hemispheres of females, in general, and thus reduces the likelihood they will develop the intellectual and emotional "disconnects" that characterize AS. (An example, mentioned earlier, is that sparse interhemispheric connectivity may prevent Aspies from rapidly grasping or producing tone-of-voice nuances produced by the right side of the brain that typically accompany left-hemisphere speech.) Although it is speculative, perhaps sexual dimorphism in connections through the corpus callosum contributes to the relatively low occurrence of AS in females.

To summarize, Aspie males and females differ from neurotypicals in some shared ways, but they also differ from each other. Although the general pattern of sexual dimorphism described above for AS is not a simple extension of that seen in TD individuals, there is some overlap between dimorphism in the two groups, especially when it comes to preferred interests and nurturing. Further, these shared dimorphic traits reflect trends that appear in many cultures, likely because they are hand-me-downs from our early ancestors. In other words, once again AS makes sense when seen from an evolutionary perspective. Evolution is based on the inheritance of genetically based features, of course, and we will turn to the intriguing topic of the genetic substrates of AS in the next chapter. Before doing that, however, a word from Eve on her thoughts about sex differences in autism.

EVE: AUTISTIC BOYS AND GIRLS

I think that gender is what you make of it. I mean, really, if you're going to say that a certain grouping of school subjects is meant for only one gender then you must be stuck in the Dark Ages or in support of segregation or something equally outdated in this modern age. Typically developing boys and girls tend to be interested in all types of subjects in schoolwork. I've known TD girls who were good with science and math as well as boys who were good with English. Just automatically thinking that girls will only be interested in arty subjects and that boys will be interested in subjects that are more logical isn't a correct viewpoint in my opinion, mostly because it marginalizes both genders, and that isn't something I'm comfortable with.

That said, to my knowledge Tony Attwood may have gotten it mostly right about the interests of Aspie boys. One of the boys I met at the school I went to in England for kids with autism and Asperger's was an obsessive fan about the kids cartoon show *Thomas the Tank Engine*. He brought figurines of the characters to school with him and could name each one, while another boy did science and philosophy and even founded a Facebook group for his philosophy interests. One of that boy's friends liked philosophy as well, but also liked mathematics. He helped me out with my work for mathematics class in the mainstream wing of our school. Yet another boy loved to imitate the catchphrases and favorite lines of the villains from the popular British television show *Doctor Who*. The same kid was also a big fan of Yu-Gi-Oh cards and walked around with a deck of them at recess.

I'm not sure whether Attwood got it completely right about Aspie girls. This is mostly because I haven't had much chance to get to know other girls with AS. Including me, there were only six girls attending the school, and I do not know what kind of autism each girl had. As far as I could tell, our interests were varied. One girl loved to dance, while another was a chatterer and spent most of her time talking to people or singing. Another girl who never spoke was good at swimming. Compared to TD girls my female friends at the school were easy to handle and they didn't bore me with chatter about makeup and clothing. They also tended to be as passionate about their interests as I was about mine. The girl who loved to dance would always try to get me to join her in dancing sessions, and she taught me a very theatrical way of bowing. We also created a new way of greeting each other ("Tinsel turkey chicken carrot fish cluck cluck") that no TD girl would have used for fear of embarrassing herself or causing some social faux pas. I know that makes no sense, but it was fun.

There were fifteen or so boys at the school, and I would say that they were much less well behaved in the classroom than the girls. Some of them had quite the tempers, although they were young enough at the time that one could forgive some immaturity. Boys were put in the relaxation or isolation rooms much more than we girls, although some of them were actually nice and didn't get violent. In comparison to the male Aspies, I might as well have had the words "controlled" and "calm" tattooed on me. As for Aspie boys at the university level, I'm not sure I can pass judgement considering that I have met very few. I have, however, run into two of the boys from my earlier school days in the street recently, and I have to say they seemed much more mature emotionally.

I think the females at the school were much sweeter, more sensitive, and friendlier than the boys. In comparison to the other girls, I was a little more prone to emotional outbursts, and more likely to socialise because I spent a lot of my time preparing for my GCSEs [General Certificate of Secondary Education in specific subjects] in the main part of the school where the "normal" students took their lessons.

In addition to attending the school, I was also a member of a disabled theatre group. I think that female Aspies have a propensity toward being good at drama or acting, perhaps because it gives us a chance to act like we don't have the condition on occasion, as Tony Attwood has suggested. The fun thing about acting is that with a script you know exactly what to do and have a small chance for making your own decisions as to some of what your character is meant to feel or act. Our group did a performance called *Forgotten* with another theatre group. It started out as a story my group made up in our sessions and ended up as a sponsored performance. Anyway, the upshot of the entire production was we raised some awareness of what happened in mental asylums back a few years ago. Although I think that girls with Asperger syndrome tend to be good actresses, I don't think I could make other people think I didn't have what I have.

Attwood's other views on Aspie girls apply only partly to me. I do like books, but classical literature that doesn't have some elements of fantasy or mysticism bores me, and I don't tend to like only one particular type of animal. I am interested in some animals but I hate bugs and animals that are dangerous or have loud calls. My favorite animals are cats and dogs as long as they are friendly.

In my opinion girls with AS may be especially fond of animals because animals don't seem to demand a series of Herculean or Sisyphean tasks from us. They don't expect us to talk to them or look at them. We don't even have to really get dressed up for them. All you have to do for animals is feed them, brush them possibly, give them water, and in the case of dogs walk them. Animals don't expect us to have to work hard to be their friends. All we have to do is take care of them, which is sort of a change from others taking care of us. It might actually be, I guess, a way of showing trust in the person with Asperger syndrome. I may not be able to cook with anything but a microwave or a toaster, but I can actually care for an animal. I think our liking for animals is actually stronger than other people's. My sister, Helen, owned a chinchilla she named Pogo. I thought Pogo was one of the cutest things I had ever seen. It helped that chinchillas have soft fur and I really enjoy the feel of soft things apart from sand, which eventually starts to feel itchy. We Aspies pay attention to animals around us while we ignore humans around us. I have to say that I played more with the white rat I owned as a teenager than I did with the other kids my age. I even named the rat after a boy I kind of liked in school.

Geeks, Genes, and the Evolution of Asperger Syndrome

In many cases the ancestors of these children have been intellectuals for several generations and have been driven into the professions by their nature. Occasionally, we found among these children descendants of important artistic and scholarly families.

—HANS ASPERGER[1]

I n his 1859 treatise on the theory of evolution, British naturalist Charles Darwin explained why traits that pass down through generations of plants and animals change over time (called "descent with modification").[2] Twelve years later, Darwin applied his theory to humans.[3] Although he was keenly aware that babies resembled their parents, he did not know about genes or their role in heredity. At the same time Darwin was pondering human evolution, however, Augustinian friar Gregor Mendel was busy discovering the fundamentals of genetic inheritance by crossbreeding different varieties of pea plants in his monastery's garden in what is now the Czech Republic.[4] Because Mendel was relatively unknown, it would be more than half a century before his discoveries about genetic inheritance were combined with Darwin's theory of natural selection in what has been dubbed the "Modern Synthesis." Thanks to this synthesis, we now know that genetic variation is at the heart of evolution. If we are to consider Asperger syndrome from an evolutionary perspective, we must therefore explore its genetic underpinnings.

Mendel's Epeaphany

Mendel patiently studied how seven features of garden pea plants were transmitted from generation to generation by crossing different varieties. Each feature came in two forms, such as tall or short plant height, round or wrinkled seeds, white or purple flowers, and yellow or green pea pods. By crossing and growing many thousands of plants, Mendel learned not only that inheritance of each characteristic was governed by the same rules but also what those rules were.

Take height, for example. When Mendel crossed pure varieties of tall and short pea plants, the resulting plants were all tall. When he crossed these offspring with each other, however, three out of every four offspring were tall, but one was short. From such experiments, Mendel reasoned that, for each trait, one discrete physical unit (known today as a gene) passed from each parent to the offspring and that these two units together determined the offspring's appearance. He also learned about the interactions and expressions of what we now call dominant and recessive forms of individual genes, figured out that each trait was separately inherited, and inferred the elegant and simple mathematical rules of genetic inheritance.[5] No wonder he is known as the father of modern genetics!

Since Mendel's time scientists have learned a good deal more about genetics. In 1953, James Watson and Francis Crick discovered that DNA had the structure of a double helix, which eventually allowed scientists to decipher how genes (which are made of DNA) manufacture the ingredients (proteins) that are combined to produce the traits of various organisms.[6] It is now known that humans have around twenty to twenty-five thousand genes distributed across forty-six chromosomes (twenty-three from each parent) that are located in each cell of their bodies, except sperm and egg cells, each of which has twenty-three chromosomes. These genes comprise only a small portion of all human DNA, however. The majority of DNA regulates the functions of the "coding" genes that, ultimately, generate proteins.

DNA is made up of long strands of building blocks called nucleotides, of which there are four kinds. Somewhat like an alphabet, the sequence of nucleotides determines the particular protein produced by a gene. This is important because our brains and bodies are composed mostly of protein. If even one of these nucleotides changes (a mutation), it may or may not change what the gene does. Mutations may also involve more of the DNA, or even whole chromosomes. Although some mutations have terrible consequences for organisms, most are neutral and some even have positive effects. In fact, mutations are the *only* source of entirely new genetic variation and thus may benefit a species if its environment changes[7] or contribute to the emergence of new species. Mutations are, therefore, natural genetic variations that have extremely diverse potential outcomes. Despite the fact that the term "mutation" has sometimes been associated with stigma (when discussing humans), such a connotation would be misguided and inconsistent with the evolutionary perspective of this book.

Human genetics are *way* more complex than those of Mendel's pea plants. Returning to the example of stature, although height in pea plants is determined by two possible kinds of one gene, a recent study of over a quarter million people revealed that adult height is influenced by variations found in more than four hundred different genes across the entire sample.[8] (These genes were not just for stature; they also influence other traits.) Further, unlike simple Mendelian traits, stature in humans occurs in all gradations between short and tall.

This is not to say that one's stature is determined only by genes. Environmental factors such as nutrition also contribute to stature, just as slight variations in soil and water conditions would have caused variation in stature among Mendel's pure strains of tall and short pea plants. Geneticists can estimate the relative contributions of genetic and environmental influences for human traits by studying them in identical versus fraternal twins. From such studies, we know that stature is highly heritable in people. In other words, despite environmental influences, a great deal of variation in height is due to genes. The same is true for most human traits, including intelligence, autism, and AS. The high heritability and extensive genetic variation described below for AS make it particularly suitable for exploring its evolutionary roots.

Before we do that, however, it is important to clarify what, exactly, is meant by "evolution." Thanks to Darwin, many people equate evolution with the origin of new species. Although speciation, indeed, characterizes one form of evolution (called macroevolution), most evolution does not produce new species. Instead, small changes can occur in the relative number of particular variants (or alleles) of specific genes from generation to generation within a single population, causing relatively minor changes in its genetic makeup, a process known as microevolution. For example, the proportions of the alleles for blood types A, B, and O might alter slightly from one generation to the next in a given human population. Macroevolution, on the other hand, refers to larger-scale changes that affect populations *as a whole*, which leads to the formation of new species.[9]

A classic example of microevolution occurred thousands of years ago when humans introduced agriculture into equatorial Africa. Curiously, this new cultural practice was associated with an increase in the sickle cell variant (allele S) of a specific hemoglobin gene relative to the gene's normal allele (A) in the population. This was unexpected because S causes death from sickle cell anemia if one has two copies (one from each parent). Such microevolutionary shifts in the frequencies of alleles may be caused by new mutations, random chance (genetic drift), migration of individuals between populations (gene flow), or, if the shifts have an evolutionary advantage or disadvantage, natural selection,[10] which are the four classic forces of evolution.

In the case of the hemoglobin gene, S increased at the expense of A despite being harmful in double dose because having only one copy of it conferred an evolutionary advantage against malaria, which had become more prevalent with the development of agriculture.[11] (That is, malaria was more likely to occur in individuals who carried two As than in those who had one A and one S.) Microevolutionary changes of the hemoglobin gene stopped in this population, however, when the relative amounts of S and A became sustained, or balanced, from one generation to the next. The sickle cell story shows that the timing and outcome of microevolutionary processes is, at best, erratic. Put more eloquently, "Macroevolutionary transitions may ultimately arise through microevolution occasionally 'writ large' but are perhaps temporally characterized by microevolution 'writ in fits and starts.'"[12]

Geneticists refer to conditions, like that for hemoglobin, in which two versions of a gene are maintained in a population because individuals who carry one of each are more likely to survive than people who have two copies of either variant as "balanced polymorphisms." The take-home message is that genes that have some negative effects can increase and be maintained in future generations because they also have positive effects when combined with certain other genes. As we will see, this basic concept is important for this book because it may partly explain the evolutionary staying power of AS.

Nature, Nurture, and Autism

It is well known that the prevalence of autism has risen dramatically in the past decade. In 2010, 1 in 68 of 363,749 (about 1.5 percent) eight-year-olds from sites in eleven U.S. states were diagnosed with Autism Spectrum Disorder,[13] and by 2012 an occurrence of 2 percent was reported among U.S. children between the ages of six and seventeen, representing "a significant increase."[14] With respect to AS specifically, the 2010 study of eight-year-olds found that, on average, about 11 percent of the 3,822 individuals who were diagnosed with specific subtypes of autism in 2010 had AS. Discussion about the global prevalence of autism and AS is postponed until the next chapter.

Meanwhile, not everyone thinks that all the apparent increase in autism is real. Instead, many believe that most of the rising prevalence should be attributed to increased awareness and acceptance of the condition, as well as to modified definitions and improved diagnostic methods.[15] Although these factors have undoubtedly contributed to the recent rise in autism diagnoses, mounting evidence confirms that at least some of the apparent increase in autism is, in fact, real[16] and, further, that the rise may be more prominent at the milder, high-functioning end of the spectrum.[17] One recent analysis, for example, estimates that about one-half to two-thirds of the increase in diagnoses of autism "reflects an increase in the true prevalence of the disorder."[18]

Although assessing the relative importance of genetic and environmental variables for the different kinds of autism is tricky,[19] tried and true methods exist for doing so. One can compare autism's frequency in identical twins (who share the same genes) and fraternal twins (who, like siblings, have only half their genes in common). These studies allow researchers to calculate the heritability of autism and thus measure the proportion of its variation that is due to genetic rather than environmental differences. Autism spectrum disorders are among the most heritable of all mental disorders,[20] as confirmed by many twin studies that quantify the strong genetic effects of its subtypes, including AS.[21] The high heritability of autism, of course, helps explain the well-known observation that neurotypical relatives of autistic individuals often have elevated levels of autistic traits.[22]

Environmental Influences Also Contribute to Autism

Despite the fact that autism is highly heritable, its causes are extremely varied and, of course, involve interplay between genetic and environmental factors.[23] Before twin studies in the last part of the twentieth century revealed that its causes are predominantly genetic,[24] autism was attributed mainly to environmental factors such as emotionally cold ("refrigerator") mothers.[25] Although this idea has been debunked, other environmental circumstances are now known to increase the risk that infants will eventually be diagnosed as autistic. These include advanced age of parents at conception (including fathers[26] and mothers[27]), exposure of gestating mothers and fetuses to various chemicals, malnutrition in gestating mothers, and birth complications.[28] Interestingly, supplements of folic acid for mothers in early pregnancy may, to some degree, be protective against autism.[29]

Another idea that has been thoroughly refuted is that vaccinating infants increases their chances of developing autism.[30] This alarmist claim originated in a 1998 British study that asserted a link between the measles, mumps, and rubella (MMR) vaccine and autism. In the wake of the study's publication in a widely respected journal (*Lancet*), vaccination rates decreased while the diseases they prevented increased. Despite the fact that the study has now been resoundingly dismissed as fraudulent in the face of massive long-term studies and the publication officially withdrawn, not enough parents appear to have gotten the message that vaccinating their children is not only safe but prudent.[31] For example, a 2016 study of more than ten thousand children in the United States, half of whom had an older sibling with ASD and half who had an older sibling without ASD, found that the younger siblings of autistic children were significantly less likely to be vaccinated than the younger siblings of unaffected children, not only for MMR but also for influenza, hepatitis, and chicken pox.[32] Not being vaccinated, of course, increases children's chances of getting these diseases but does nothing to reduce their risks for developing autism. Appropriately, the authors of the study concluded that public health officials and clinicians should do more to inform parents of children with ASD about vaccines.

Although a few studies have concluded that susceptibility to autism is substantially influenced by environmental factors,[33] recent analyses of many earlier twin studies show that the bulk of the evidence points to much stronger genetic factors.[34] For example, a study of all the twins born in England and Wales over a period of three years found that autism, including forms on the high end of the spectrum, was the result of "additive genetic influences and a smaller proportion attributed to non-shared environmental influences. . . . We found very little evidence of shared environmental effects overall."[35]

Few, if any, studies distinguish the environmental risks associated with AS from those for autism in general. One recent suggestive finding, however, was that a combination of maternal prepregnancy obesity and diabetes was associated with a significant risk for bearing infants who would eventually be diagnosed with both autism and intellectual disabilities. This finding led the authors to conclude that the causes of severe autism may differ from those of the less severe forms.[36] Because the latter category is comprised largely of Aspies and people with high-functioning autism, the prenatal environmental impact of maternal obesity and diabetes may not be as significant for these types of autism as for other forms. More broadly, it suggests that the extent and type of environmental factors that impact autism may vary somewhat for its different subtypes—more research is needed.

Although there is less uncertainty about the overall impact of genetic factors on autism, the specific genes that contribute to it are just beginning to be unraveled and are incredibly complex,[37] with estimates of up to one thousand genes implicated as contributing to different cases of autism.[38] Further complicating the matter, "no specific gene accounts for the majority of ASD; rather, even the most common genetic forms account for not more than 1–2% of cases."[39]

The Genetic Underpinnings of Asperger Syndrome

In 1944, Hans Asperger was the first to suggest a strong genetic component for AS, when he observed that, among more than two hundred children with autism, "we have been able to discern related incipient traits in parents or relatives in *every* single case where it was possible for us to make a closer acquaintance."[40] Since then, Asperger's observation has been borne out by various reports of autistic traits in family members of Aspies, particularly among fathers.[41] It is also consistent with the finding that three times as many parents and grandparents of individuals with AS have autistic behaviors as compared to the parents and grandparents of people with HFA.[42]

Happily, scientists are beginning to gain some insight into the genetics of AS. One study, for example, compared the building blocks, or nucleotides, of the DNA in all the chromosomes of affected and unaffected individuals in over two hundred families that had at least one Aspie member, with fascinating results.[43] The authors identified the top one hundred locations across the full set of chromosomes where specific nucleotides appeared more often in Aspies than their unaffected relatives. All in all, twenty-six different genes on nineteen of the twenty-three chromosomes were found to be vulnerable to one or more of these tiny variations in Aspies, and eight of these genes are known to be highly important for the brain.

Most of the Aspie's unusual variations, however, were found in parts of chromosomes that do not contain genes but instead regulate the activity of genes. As one would expect given overlapping diagnostic criteria, some of the variations associated with AS have also been reported for ASD, including one on a bit of chromosome 7 that is known as the susceptibility locus gene (*AUTS1*), which supports the suggestion that various kinds of autism share some relatively common genetic features.[44] (Common genetic variants are typically defined as those that occur in more than 5 percent of the general population, whereas rare ones occur in less than 5 percent,[45] although some researchers use a different percentage as the cutoff point.) The authors are careful to point out that their findings also suggest that certain genetic risks may be unique to AS.

Another method for detecting the genetic substrates of AS focuses on common genetic variations within genes that have specific known functions in the general population, and that might conceivably influence certain Aspie traits. For example, one study that examined variations in a large number of preselected genes that occurred in at least 20 percent of the general population detected six unusual variations associated with AS—three of these so-called candidate genes are known to have functions related to sex hormones, two with neural connectivity, and one with social emotional responsiveness.[46]

In addition to relatively common variations in tiny parts of (or between) certain genes, much rarer mutations have also been associated with AS. In these cases, the mutation often occurs spontaneously in the individual or in one of his or her parents (so-called *de novo* mutations).[47] *De novo* mutations may also be caused by deletions or duplications in larger segments of chromosomes (copy-number variations, or CNVs), and these kinds of mutations have been reported for individuals with AS.[48] For example, a deletion of some twenty-seven genes on chromosome 20, including the oxytocin gene (*OXT*), has been found in one boy with AS.[49] Despite the fact that this result was for just one boy (thus needing replication in much larger samples), a subsequent study that included 174 individuals with AS confirmed the association of a common genetic variation within *OXT* with AS.[50] More recently, a common variant in the oxytocin receptor gene (*OXTR*) has also been found to be strongly associated with AS in a study that included 118 cases of AS and 412 controls.[51] These findings are particularly relevant with respect to the evolutionary trends discussed in chapter 1 because oxytocin is linked to (among other things) various aspects of socialization, including parent-infant bonding in humans and other mammals.[52]

Despite the *DSM-5*'s removal of Asperger's Disorder as a discrete form of autism, geneticists continue to search for and/or confirm variations in genes that are associated with it. And the list is growing.[53] Although it should be kept in mind that "contributions from rare and common genetic variants are not mutually exclusive," most of the genetic variations that have been linked to AS are not rare.[54] (It should be noted, however, that recent research suggests that rare *de novo* mutations contribute to at least 30 percent of autism in families that have only one affected member.)[55] It thus appears that the additive influence of many common variants within an individual, each with a very small effect, has a relatively large impact on the potential risk of developing AS,[56] compared to the risk from *de novo* and other rare mutations, as is the case for autism generally.[57] More specifically, one recent review estimates that over half of the families in which more than one individual has autism may have several common alleles that, when combined in a prenatal infant, have an additive impact on the likelihood of that infant developing autism.[58] What this boils down to is that, compared to Mendel's peas, the genetic underpinnings of AS are varied and incredibly difficult to decipher.

But scientists are making progress. As we saw in chapter 1, AS is basically neurodevelopmental in origin and one of its most distinctive features is a tendency for infants to experience relatively accelerated brain spurts early in life (the third evo-devo trend), resulting in comparatively larger brains during their first years, on average, compared to neurotypical children (fig. 1.5). As discussed earlier, this overgrowth in brain size has been attributed to an overabundance of neurons in the gray matter of the cerebral cortex and, to some extent, the white matter (myelin) that insulates nerve fibers, which are likely due to alterations in both prenatal and postnatal brain development. Enlarged brains associated with autism were also speculated to involve a reduction in the pruning of synapses compared to TD infants.[59] It is not surprising, then, that a number of the studies referred to in this chapter found that specific genetic variants associated with autism were involved in neural development and connectivity.[60] In keeping with this, recent research has shown that many genes that have been associated with ASD with high confidence are activated (coexpressed) "in mid-fetal human cortical neurons and converge on synaptic development pathways. . . . Many ASD genes showed coordinated expression in postmitotic neurons prenatally" and postnatally.[61]

Similarly, because AS is manifested differently in males and females but is associated with impaired social skills in both, the discovery that Aspies have higher frequencies than the general population of certain genetic variations that impact sex hormones as well as others that influence social development (such as *OXT* and *OXTR*, discussed above) is also not surprising.[62] Despite this progress in uncovering genetic variants that are associated with AS, it is important to keep in mind that there is no one specific Aspie gene. Thus, as is the case for autism generally, even the most common genetic variants that have now been associated with AS account for only a tiny proportion of all cases.[63]

As discussed in chapter 1, the emergence and development of language was likely the most important factor that drove brain evolution in our ancestors, which highlights the importance of investigating and comparing the genetic substrates of Aspies (whose development of language is not delayed) with those of people with HFA (whose acquisition of language is, by definition, delayed). Fortunately, a few scientists have twigged to the fact that "AS is a subset of ASC [Autism Spectrum conditions] where individuals have no language delay, suggesting it may have a genetic architecture distinct from the rest of ASC."[64] Hopefully, researchers will pursue this line of thinking in the future. For example, an exciting focus for comparative studies would be to investigate the genetic underpinnings of the intellectual strengths that frequently occur in Aspies, including an encyclopedic knowledge about preferred subjects, prodigious memory, and superb analytical skills. In a similar vein, it would be stimulating to explore the genetic substrates of the extraordinary visual thinking associated with HFA. In fact, these are such potentially productive avenues of research that one wonders why they have not yet been pursued.

AS: A Genetic Paradox or Not?

Let me be clear. I do not mean to suggest that, during human evolution, natural selection positively targeted AS (or HFA) in and of itself. As we have seen, despite their great systemizing and other intellectual strengths, Aspies have severe deficits when it comes to interacting socially. Furthermore, these deficits are associated with serious consequences. As with other forms of autism, individuals with AS are almost twice as likely as TD individuals to die before they reach their early forties, frequently because of accidents, suicide, or diseases that are sometimes associated with AS, such as epilepsy.[65] Nor are Aspies as likely to marry (and, presumably, have children) as neurotypical individuals, although they seem to do so more than other persons with autism.[66] Stated in evolutionary terms, this means that Aspies have reduced reproductive fitness compared to the general population—that is, as a group Aspies do not make as large a genetic contribution to future generations.

Put in even starker terms, people with autism including AS are being selected against.[67] Why, then, is the heritability of AS so high? How can the genes for a condition that is at a strong reproductive disadvantage contribute so much to that condition in future generations? In other words, since Aspies provide relatively fewer genes to future generations, shouldn't natural selection be removing their genes from our species' genetic repertoire (or so-called gene pool)? This apparent paradox has many geneticists and evolutionary biologists scratching their heads because the prevalence of AS is not only failing to decrease over time, it is clearly increasing in at least some communities (as discussed in the next chapter).

Although scientists are divided in their explanations for this conundrum, they generally acknowledge that autism, including AS, is an incredibly complex condition that is influenced by a combination of environmental and genetic factors that can include a huge number of genes as well as other noncoding regions of DNA that boss genes around—but with different combinations in different individuals.[68] Despite the fact that researchers disagree about the relative contributions of rare versus more common mutations when it comes to explaining the ongoing prevalence of autism, most accept the idea that some of the common alleles that contribute to it may have positive spinoffs that keep them in the gene pool, as discussed above in regard to sickle cell anemia.[69] In the final analysis, and despite quibbles about their respective contributions, both "large-effect rare mutations and small-effect common variants contribute to risk" for developing autism.[70]

It is well established that autism is the result of many genes acting together, as is the case with respect to other complex traits such as stature and IQ. Psychologists Annemie Ploeger and Frietson Galis of the University of Amsterdam reason that some of the genes underlying autism may also contribute to other traits that benefit affected individuals. If so, the multiple effects of genes could help resolve the apparent paradox of high heritability but low reproductive fitness in autism:

> For example, it is possible that genes involved in the development of autism are also involved in the development of intelligence. As intelligence is positively correlated with reproductive success, genes involved in autism can possibly spread in the population. We propose that in most individuals, the interactions between genes result in normal or high intelligence and the absence of autism. However, in some unlucky situations, often in combination with spontaneous negative mutations, the interactions between genes can lead to the development of autism. . . . Thus, the combination of high heritability and low reproductive success in autism can be explained from an evolutionary developmental perspective.[71]

Whether or not one characterizes autism as "unlucky," which is debatable,[72] one wonders what kinds of cultural, genetic and, ultimately, environmental factors might account specifically for the emergence of AS in individuals, as well as its persistence across generations.

The Geek Hypothesis

British psychologist Simon Baron-Cohen theorizes that, although everybody has some ability to understand predictable abstract patterns, people with autism tend to be "hyper-systemizers" who can "only process highly systemizable (law-governed) information."[73] Although this might be a bit of an overstatement, Baron-Cohen and his colleagues have, indeed, amassed a good deal of evidence linking autism and systemizing abilities. Examples include the discoveries that the parents of Aspies test relatively high for systemizing traits,[74] that there are more engineers among the relatives of autistic children,[75] and that autism is highly prevalent among math majors and their relatives.[76]

Building on these findings, Baron-Cohen famously, if controversially,[77] argues that AS is rising especially rapidly in technologically inclined communities like Silicon Valley, California,[78] "Route 128 outside Boston,"[79] and the Eindhoven region of the Netherlands.[80] According to this hypothesis, analytically inclined men and women (systemizers, as Baron-Cohen calls them) are drawn to technological enclaves where they meet, marry, and have children who are likely to inherit double doses of some of the common genetic variations that underpin their parents' technical skills. The tendency for geeks to meet and marry is an example of what biologists call positive assortative mating, in which individuals with similar genetic or physical makeup mate with each other more often than expected by chance ("likes mate with likes"). Baron-Cohen's suggestion, which recalls Hans Asperger's quip that a "dash of autism" is essential for success in science, as well as his observation about the intellectual and artistic lineages of Aspies quoted in the epigraph for this chapter,[81] is currently finding support from other researchers.[82] (Eve [fig. 5.1] fits the offspring-of-geek profile, with her mathematician father and lawyer mother.)

Although the geek hypothesis is frequently alluded to in conjunction with AS,[83] the extent to which it applies to individuals with HFA is an open question.[84] This lack of clarity is likely due to the fact that HFA is imprecisely and broadly defined compared to AS and has not been as thoroughly studied in its own right. Nonetheless, certain anecdotal information suggests that some of the experts in technical enclaves may have HFA, especially if HFA is associated with relatively enhanced right-hemisphere visuospatial skills, as theorized in earlier chapters. For example, referring to autistic employees at Microsoft, a supervisor there told science writer Steve Silberman that "all of my top debuggers . . . can hold hundreds of lines of code in their head as a visual image. They look for the flaws in the pattern, and that's where the bugs are."[85]

The strongest evidence that I am aware of that supports a probable link between positive assortative mating and a rise in the prevalence of AS comes from University of Chicago–trained economist Hays Golden.[86] Golden developed a mathematical model of genes and assortative mating, which he applied to five years of data (2000–2008) from the Centers for Disease Control and Prevention (CDC). He not only concluded that autism is connected to systemizing (gleaned partly from analyzing data about occupations) but also provided persuasive evidence that "assortative mating could be playing a significant role in the rise of autism rates, and more generally, that even in cases where assortative mating does not dramatically change the population distribution, small populations at the extremes may be dramatically affected."[87]

Golden theorizes that if assortative mating has significantly contributed to the rise in autism, it is likely to continue rising for several more generations

Figure 5.1a and 5.1b. Girls can be geeks, too. According to the "geek hypothesis," analytically inclined parents are likely to transmit more genes that underpin systemizing abilities to their children than less analytically inclined parents. Illustration reproduced with permission of Pete Ellis; photograph of Eve as a little girl provided by her mother, Sarah Nichols.

because it takes time to reach a new genetic equilibrium. Thus "investments in better autism care are probably more cost-effective than they appear at the current prevalence."[88] Reassuringly, Golden ventures that increased assortative mating between systemizers "has probably been socially beneficial, despite the increase in autism rates."[89] It is thought provoking to contemplate that, thanks to advances in transportation, education, and technology, opportunities for assortative mating are likely to continue to increase:

> My model suggests that developments such as computers, which have likely increased the returns to [financial rewards for] systemizing, have led to more assortative mating on systemizing. The same is true for the shifts that have caused women to enter systemizing occupations at much higher rates. I find evidence that mating on observable systemizing has increased, and that the increase reflects a true shift in who marries whom.[90]

Some might argue that a portion of the relatively greater incidence of AS (and probably HFA) in places like Silicon Valley is due to various factors such as modified definitions and improved diagnostic methods as well as enhanced awareness and acceptance of autism, which account partly for the rise in contemporary compared to past headcounts for ASD. Although it is difficult to understand why health professionals in technological enclaves would use definitions and diagnostic criteria that differ from those of diagnosticians elsewhere, it is entirely possible that awareness of autism could be enhanced in the general population of technological communities, and that being an Aspie or having HFA might even be associated with a certain cachet in places that value systematizing skills, similar to the "cool" attributed by some to the "geek chic" style. If so, autistic individuals might seek confirmatory diagnoses, which could contribute to the relatively high frequencies observed for AS/HFA in technological compared to other communities. It is also worth noting that parents sometimes realize their own symptoms when their children receive a diagnosis of ASD and may seek a diagnosis for themselves, or providers may offer to evaluate them when they examine their children.

Meanwhile, if the frequencies of AS and HFA are, indeed, increasing more in the Silicon Valleys of the world than in other communities because of positive assortative mating (which seems likely based on the reports discussed above), it is not only an instance of microevolution in these places, it is also an example of sexual selection, which Charles Darwin recognized long ago as fundamentally important for human evolution.[91] What is particularly appealing about the geek hypothesis is that it recognizes that the genetic underpinnings of AS/HFA contribute to intellectual *strengths* in addition to deficits.[92] This is refreshing because clinicians tend to pathologize the positive aspects of AS, for example by characterizing the extraordinary vocabularies of some Aspies as "pedantic" or the expert knowledge many have acquired as spinoffs of "obsessive repetitive behavior."

Aspies in Prehistory?

In addition to positive assortative mating in Silicon Valley and elsewhere in the world, prehistoric evolutionary factors must have influenced the emergence and subsequent retention of AS in the human lineage because it occurs only in people. It is becoming possible to obtain hints about how this might have happened by carrying out DNA studies on material harvested from the fossilized bones and teeth of some of our ancestors, such as Neandertals and their close cousins, the Denisovans.[93] Such studies have recently identified specific genes in humans that contain novel mutations that did not exist in their archaic cousins. These new variations became fixed (i.e., occur in 100 percent of the population), or nearly so, in the human lineage *after* our ancestors diverged from these archaic relatives.[94] Many of these variations appear to have been advantageous ones that involved aspects of skin pigment, immune responses, and the brain.[95]

One of the most dramatic examples of a newly mutated region in humans involves chromosome 16. Unlike other primates, including hominins such as Neandertals, the human version of this chromosome contains one or more extra copies of a segment that contains a gene called *BOLA2* because of a mutation that occurred about 282,000 years ago, about the time *Homo sapiens* first emerged. Intriguingly, "recurrent copy number variation (CNV) at chromosome 16p11.2 [the position on the chromosome] accounts for approximately 1 percent of cases of autism and is mediated by a complex set of segmental duplications, many of which arose recently during human evolution."[96] Further, these duplications contributed "more derived sequence specific to *H. sapiens* than 35,500 previously reported human-specific single-nucleotide variants [mutations of DNA bases] and indels [insertions or deletions of DNA bases] combined. The rapid rise and dispersal of this duplicated segment at the root of *H. sapiens* . . . are unlikely to have occurred under neutral evolution but rather are consistent with modest positive selection."[97] Although *BOLA2*'s function is still not well understood (it likely plays a role during human embryonic development), selection for duplications in this segment of chromosome 16 in humans appears to have predisposed a "critical region" of the chromosome that is flanked on either side by duplications of *BOLA2* to *other* variations that are associated with autism.[98] This suggests that some of the susceptibility to autism in humans arrived on the coattails of positive selection for *BOLA2*.

Another region that differs noticeably between living people and Neandertals is part of a gene on chromosome 7.[99] Although the functions of the gene, called *AUTS2*, are not well understood, it appears to play a significant role in neurodevelopment. It is not surprising, therefore, that it seems to have been important for human evolution. What is particularly noteworthy, however, is that specific disruptions in *AUTS2* have been recorded for more than thirty individuals with autism.[100] Another gene (*KLHDC10*), which codes for a protein expressed in adult and fetal brains, has two common variants that have been associated with AS. Tantalizingly, this gene is located in the same general neck of the woods as *AUTS2* on chromosome 7.[101]

The gene known as *FOXP2* is also located near the *AUTS2* gene on chromosome 7. Since diverging from chimpanzees, human ancestors underwent two tiny mutations in this gene, which produces a protein that regulates other genes.[102] Unlike the cases we discussed above, however, these mutations occurred before rather than after humans and Neandertals diverged. This explains why both species have the same two evolutionary changes. Although other forms of *FOXP2* occur in many animals and have many functions, its evolved form in humans contributes to individuals' development of language and speech, which suggests that alterations in this gene may have facilitated the emergence of human language. Since language impairments frequently accompany autism, some researchers have suggested that common variants of *FOXP2* might be involved in autism. However, a thorough analysis of the *FOXP2* gene in 322 unrelated individuals with autism, including some Aspies, proved that this is not the case, although rare variants of this gene have occasionally been associated with autism.[103]

Despite the finding that common mutations in *FOXP2* are not associated with AS, we should keep in mind the potentially transformative effects that initially rare genetic mutations might have had during early hominin evolution if they happened to become a target of natural selection and, thus, eventually became the norm. Given what we know about *FOXP2*'s role in the acquisition and fluency of speech in humans, it is easy to imagine that the pre-Neandertal hominins in whom the mutations first arose developed an odd habit of chopping up little bits of air and spewing them from the mouth in an effort to communicate with others. Those ancestors may well have been the first geeks (or, at least, the first proto-geeks). In fact, what may once have been a prehistoric developmental anomaly (or "syndrome") has, over time, become universal and is probably the most dramatic illustration of what differentiates us from all other primates. Without speech, what would we be?

Although ancient DNA studies are in their infancy, the few results discussed above suggest that at least some of the common mutations that have been implicated in autism may have emerged as variations within genes that were under intense positive selection after the human lineage diverged from earlier ancestors.[104] In particular, relatively common genetic mutations that likely underpin the cognitive strengths of Aspies may have been swept along on the evolutionary "coattails" of natural selection for advanced cognition in our species. As science journalist Steve Silberman put it:

> For all we know, the first tools on earth might have been developed by a loner sitting at the back of the cave, chipping at thousands of rocks to find the one that made the sharpest spear, while the neurotypicals chattered away in the firelight. Perhaps certain arcane systems of logic, mathematics, music, and stories—particularly remote and fantastic ones—have been passed down . . . , in parallel with the DNA that helped shape minds which would know exactly what to do with these strange and elegant creations.[105]

Furthermore, if Baron-Cohen is right, AS is likely to continue to increase in at least some communities as *Homo sapiens* becomes more and more enthralled with information technology. (Much more about this in chapter 7.)

Hans Asperger was the first to suggest that AS has a strong genetic component when he observed that the unaffected relatives of his young patients shared many of their Aspie traits. He was right, of course, as confirmed by the recent research of Baron-Cohen and many others. Figure 5.1 nicely illustrates the idea that Aspies tend to inherit their systemizing abilities from both parents. Since Eve's background fits well with the geek hypothesis, let's see what she has to say about her family. As you will see, her comments underscore the hugely important role that siblings and parents play in the lives of Aspies.

EVE: MY FAMILY

When I was between four and fifteen years old, I lived in America during the school year with my mother, Sarah, and two younger sisters, Helen and Judith. I liked living with Mama because she was so nice to us. She made sure there was a computer I could use for playing CD-ROM games, and Internet games, as well as a TV, VCR, and DVD player when those technologies came out. This helped me find my own way of relaxing and zoning out. We also got to see Grandma Dean on a regular basis. I found out that I had Asperger syndrome at the age of nine. Although my mother told me, she wasn't the first person to think that I might have it. That particular task fell to one of my teachers in elementary school who noticed that my behavior didn't seem to be a lot like the other children's.

I didn't exactly have a lot of emotional control back then, and the sisters were the people I interacted with the most. Some of my earliest memories involve my temper or playing games with Helen and Judith. They always seemed to be more mature than I was and they acted so superior sometimes. I used to call them *Bratasaura regina* and Princess Brat because they always seemed to get their way. I am somewhat better in my behavior toward the sisters now, though I still call them a variety of names that are related to either their childhood nicknames or their actual names. Helen is Meems or Lenny, based on her current nickname of Lena, and Judith is Udie, based on the middle syllable of her name.

When my behavior tended to catch attention from other people at school, the sisters defended me by telling them that it was none of their business how strangely I acted or dressed. They also tried to get people to stop teasing me because of my stammer. I was somewhat gullible and ended up doing quite a bit of what other people told me to do. That is, on occasion I would end up singing in corridors or on the playground because other students told me that it was a good idea. Helen got somewhat run ragged trying to get people to stop trying to make a fool of her sister.

When I was fifteen, my sisters and I moved to England to live with our father, Aidan, who is a mathematician. A main reason we went to live with Dad was that the government paid for English students to attend university there. I wasn't as nervous about moving to England as I would have normally been about a new experience because the sisters and I had visited Aidan every summer after we moved to America. At the time I just thought it would be a longer version of the normal trip, and besides I have family on both sides of the Atlantic. Living with Dad was more structured than living with Mama, partly because he found a school for me that was specifically for kids with Asperger's and autism, which helped me get my life in order. Because of Aidan, I've managed to get into university and have just graduated.

As adults, my sisters and I are very different people. I'm a worrier with an unsteady optimism streak, and have a semi-held-back temper in comparison to when I was younger. Helen and Judith are a lot more independent than I could hope to be. Unlike me, the sisters are able to plan more than a day ahead, and know how to prepare for said plans. Helen has just finished a degree at the London School of Economics and is now working on a master's in social work practice at the University of York. I would describe Meem's personality as responsible, caring, and take-no-prisoners. Which is to say that my sister doesn't hold back on her opinions of other people, and will always make sure that she does what she can to help you improve. Case in point, she was elected Women's Officer for LSE's student union and did everything she could to make her university a better place for other girls to study and work. She also had her own lodgings in London. Judith is a student in English literature and creative writing at the University of Birmingham. I would describe Udie's personality as more temperamental than Helen's, and yet her care towards others is much warmer than mine. She is always willing to try and give me a push towards independence of thought and action.

I call my father a variety of different funny names that really don't make a lot of sense, such as Spurg or Aidinsky, or on occasion I call him by his first name, Aidan. I am really not that good at the type of math you have to figure out on a piece of paper, but I can visualize simple maths formulas in my head after I see the problem written down on a piece of paper. Aidan, on the other hand, is such a good mathematician that he tends to leave his written calculations strewn all over our house, though I have to say that some of those calculations are stuff he is using for the card game Bridge, which Aidan finds interesting and I find the most boring thing in the world apart from lectures. No offense meant to my university lecturers. According to my mother, Spurg seems, like me, to have a habit of talking out loud to himself. In his case he talks about mathematics, while my choice of subject matter for my talks to myself is usually the plots of different animes or different mangas. While attending uni, I lived with father in Bristol. Spurg and I get along fairly well, and he actually gets me.

Neither of the sisters have demonstrated any Aspie traits that I can think of. They do not seem to be overly obsessed with keeping to a schedule, or keeping to a plan of how they want their day to be. Neither of them has trouble with getting the meaning of figurative language or knowing when someone is joking with them. In fact, they are much better at speaking figuratively than others I have met. They also never seem to be at a loss in social situations and have plenty of friends who keep them busy. Mama has a small amount of Aspie-ish behavior in her. She can get stressed easily and also tends to be somewhat schedule oriented, but I think this has more to do with her training as a lawyer. She seems to get figurative language and to know the social rules much better than I, and has many friends. Unlike me, Mother seems to have an active sense of humor and can actually crack jokes that other people get. She tends to make me feel as though I want to take care of her, which is something of a surprise considering I don't usually try to be responsible for other people.

I consider myself closer to Helen and Judith than I am to any other people, although I am not sure of the exact roles they will play in my future. I think I could probably count on them as sounding boards for any ideas I have about my life. After all they're my sisters, and I wouldn't trust anyone else to help me without being too critical about what I've already done. I think my family members would be happy if I just had the possibility of getting a good job. They would also like me to find a career that I enjoy doing. Maybe they would also like for me to find someone outside the family that I could enjoy spending time with, and to send them occasional updates about my life so they could develop their own lives separate from me. For now, though, I am happy to have graduated from Bath Spa University.

Autism Rising

A World Perspective

> Once one has learnt to pay attention to the characteristic manifestations
> of autism, one realises that they are not at all rare in children,
> especially in their milder forms.
>
> —HANS ASPERGER[1]

A s previously noted, it appears that the prevalence of autism has risen dramatically in recent years, although some have claimed that the increase is due mostly to enhanced awareness of the condition and changing diagnostic practices.[2] But is the occurrence of autism different in developing than in industrialized countries, and is it perceived and treated similarly across cultures? And what about Asperger syndrome? Do scientists know how widespread it is compared to other forms of autism? Does it exist all over the world?

Although these questions are difficult to address and fraught with controversy, public health researchers are beginning to provide some answers. In 2012, Mayada Elsabbagh of Montreal Children's Hospital in Canada and an international group of collaborators provided the first substantive comparison of the prevalence of autism in different parts of the world, including some low- and middle-income countries.[3] Estimates of autism's occurrence were reviewed for parts of Europe, the Western Pacific and Southeast Asia, the United States, and (to a lesser extent) the eastern Mediterranean, but adequate information was not available for Africa. The authors confirmed a rise in the prevalence of autism since the 1960s, which they attributed to various factors, including the recent broadening of definitions of autism to include less severe forms, diagnostic switching to autism from other developmental disabilities, greater availability of social services, and enhanced awareness of autism spectrum disorders by both professionals and the public.

According to Elsabbagh and her colleagues, most of the studies conducted since 2000 converge on a prevalence of 1 in 161 individuals (that is, less than 1 percent of the population) as the best estimate of a global average for autism in 2012. Although the authors documented variation from region to region, they concluded that existing evidence was too limited to support the idea that prevalence of autism is impacted significantly by geographical region, ethnicity, socioeconomic factors, or culture. That said, Elsabbagh's team also acknowledged that there was too little information to detect such effects if they did exist, particularly in low-income countries.

A subsequent study by Amanda Baxter of the University of Queensland, Australia, and her colleagues estimated the 2010 prevalence of childhood autism and other ASDs (the "other" category included PDD-NOS [pervasive developmental disorder—not otherwise specified] plus AS) for regions in various parts of Asia, Australia, Europe, Latin American, North America, Oceania, and sub-Saharan Africa.[4] Despite the fact that the authors lacked sufficient data outside of high-income countries, a suggestive pattern emerged. After adjusting for methodological variations and population growth, there was no clear evidence for a change in the prevalence of autism between 1990 and 2010, nor did there appear to be much regional variation across the globe.[5] This does not mean that the occurrence of autism was inconsequential, however: "In 2010, an estimated one in 132 individuals had an ASD, with this translating to 52 million cases of ASDs . . . across the globe."[6] Although this estimate is slightly higher than that from the Elsabbagh study, it is still less than 1 percent of the population. It is necessary to take the global estimates from both studies with a grain of salt, however, not only because they lacked sufficient information from low-income countries (more about that below) but also because these analyses would likely have been on the low side because they did not capture unreported cases.

Recent statistics for autism in the United States suggest a somewhat higher prevalence. A report sponsored by the CDC analyzed information from a 2014 survey of parents of more than eleven thousand children age three to seventeen.[7] The parents were asked, "Did a doctor or health professional ever tell you that [child's name] had autism, Asperger's disorder, pervasive developmental disorder, or autism spectrum disorder?"[8] Based on the parents' answers, the report's estimated prevalence for ASD was 2.24 percent (approximately one in forty-five), which was significantly higher than the estimate of

1.25 percent (one in eighty) reported in the United States for 2011 through 2013. The authors attribute the marked increase in 2014 to a belief that the question asked of parents was more likely to capture the full population of children with autism than questions used in earlier surveys.

Whether or not one attributes the rise in reported cases of autism in the United States to a more carefully worded survey, the fact that over 2 percent of 11,000 three- to seventeen-year-olds were diagnosed at some point in their young lives with autism is remarkable. It is not unique, however, having been matched by a prevalence of 2.2 percent in 55,266 South Korean children (born from 1993 to 1999) age seven to twelve using *DSM-5* criteria,[9] and an even higher prevalence of 2.64 percent in those same children when criteria from *DSM-IV* were applied.[10] These figures are on the high end of the global prevalence estimated by another survey, which concluded that "around 1–2% of the world's population of children has ASD; however, over 85% of cases of ASD in epidemiological studies are identified from only 10% of the world's children (North America, Europe, and Japan)."[11]

To summarize, it is unmistakable that the estimated numbers of autistic individuals have risen in certain parts of the world in recent years.[12] What is less clear are the reasons for this: "Considerable controversy exists as to what is producing these increasing rates. Many attribute these increases to changes in diagnostic criteria over time, new assessment instruments, inaccurate diagnoses, utilizing different research methodologies to identify prevalence estimates, cultural differences, and increases in autism awareness. Yet, many details within these categories even further complicate the picture."[13]

More research that takes into account the many factors that potentially influence the apparent prevalence of autism is obviously needed. In fact, without such careful studies it is predicted that "the rates of ASD will continue to rise without known cause."[14]

How Prevalent Is Asperger Syndrome?

To return to a question raised earlier, how does the prevalence of AS stack up against other kinds of autism and does it occur globally? Although there is not much information that specifically addresses this question, there is some, and it is very interesting. Psychiatrist Eric Fombonne of Montreal Children's Hospital in Canada conducted one of the first studies that compared

the prevalence of AS with that of all other forms of autism combined. Fombonne's analysis included ten surveys that were published between 1998 and
2007, six in the United Kingdom and one each in Norway, Sweden, Denmark,
and Canada. Although diagnostic criteria for AS differed somewhat among
the studies,[15] its prevalence was lower than that for autism in nine of the ten
surveys. Fombonne estimated that, on average, for every case of AS there
were three or four cases of autism in these countries.[16]

A more recent study of eighteen surveys (from 1998 through 2012) from
the above countries plus Australia, Finland, Venezuela, and the Netherlands
calculated that estimates for the prevalence of AS averaged 0.12 percent, or
slightly over one-tenth of 1 percent of the population, although the estimates
varied widely.[17] Again, the rates for AS in this recent study were consistently
lower than those for Autistic Disorder, with an average of around one person with AS for every six with Autistic Disorder. (Autistic Disorder, which
was defined in *DSM-IV* as distinct from Asperger's Disorder, would today
be included in ASD.) Still other studies have estimated a higher prevalence
of AS that is over one half of 1 percent of the population.[18] Some have cau-
tioned, however, that prevalence estimates for AS may have been inflated by
inclusion of individuals with high-functioning autism, and that the diagnostic criteria and measurement techniques for distinguishing AS from Autistic
Disorder have been inconsistent, leading them to conclude that the "data on
AS are therefore of dubious quality."[19]

But all is not lost. One of the best windows into the relative prevalence of
AS, at least in the United States, comes from a CDC surveillance of 363,749
eight-year-old children in 2010 from eleven sites in different American
states.[20] One in 68 of the children (about 1.5 percent) were diagnosed with
autism. Of these, 3,822 children received diagnoses for a specific subtype of
autism, including AS diagnosed with *DSM-IV* criteria. The percentage of
children identified as Aspies ranged from 7 to 17 percent of those diagnosed
with any form of autism across the various sites, with an average of 11 percent.
This translates to a ratio of approximately one individual with AS for every
eight with non-AS diagnoses of autism (that is, one of nine).[21] Although CDC
surveillances began incorporating *DSM-5* diagnostic criteria for ASD in 2014,
they will continue to evaluate children with the *DSM-IV* criteria for autism
as well. In other words, despite the elimination of Asperger's Disorder from
DSM-5, future analyses by the CDC should shed light on the extent to which

the "real" prevalence of AS is increasing in the United States. Meanwhile, although current estimates for the prevalence of AS differ to some extent among countries, they are considerably less than 1 percent of the population, with the highest recent estimate for any country being six-tenths of 1 percent of the population (0.60 percent) for all children age seven to twelve in South Korea, which also has the highest estimated prevalence for autism generally.[22]

Are Autism and Asperger Syndrome Just Plain WEIRD?

A fascinating article titled "The Weirdest People in the World?" by Joseph Henrich, Steven Heine, and Ara Norenzayan of the University of British Columbia in Vancouver, Canada, is highly relevant for this book, although it never mentions autism.[23] The authors showed that behavioral scientists routinely publish broad claims about human psychology and nature based on studies that focus exclusively on Western, educated, industrialized, rich, and democratic (WEIRD) societies. "Western" countries include those "clustered in the northwest of Europe (the United Kingdom, France, Germany, Switzerland, the Netherlands, etc.), and British-descent societies such as the United States, Canada, New Zealand, and Australia."[24] After comparing a huge number of studies across wider areas of the world on myriad topics, including (to name just a few) visual perception, cooperation, spatial reasoning, other reasoning styles, and self-concepts, the authors concluded that "members of WEIRD societies, including young children, are among the least representative populations one could find for generalizing about humans. . . . We need to be less cavalier in addressing questions of *human* nature on the basis of data drawn from this particularly thin, and rather unusual, slice of humanity."[25] Despite the fact that, as noted, autism was not discussed in the paper, flipping back a few pages in this book will confirm that the societies that formed the basis for various estimates of the global prevalence of autism are, indeed, WEIRD ones.

Although Henrich et al.'s study confirmed the universality of some psychological features such as susceptibility to particular optical illusions and the emergence in children of an ability to pass false-belief (ToM) tests, certain unusual traits stood out in the "Western" societies. For example, compared to other cultures, people living in Western societies appeared to have relatively strong predilections for analytical versus holistic reasoning, as well as for individualistic concepts of self as opposed to thinking of themselves as being intertwined in social webs with obligations toward others. This finding for whole cultures is strikingly reminiscent of the systemizing and empathy quotients so often used to identify individuals with autism, which raises the interesting question of whether autistic people, including Aspies, represent an extreme manifestation of the Western norms for analytical thinking and social independence instead of, or in addition to, reflecting a universal human condition.

This question is easier to address for autism writ large than it is for AS. As we have seen, too few data are available to estimate the prevalence of autism in underdeveloped countries with much confidence. Nevertheless, beginning in 1978, autism has been, and continues to be, reported in countries classified as low or middle income.[26] By 2013 autism had been reported in at least forty countries with annual gross national incomes (GNI) per individual below $3,975, and of these, twelve had GNIs below $1,005.[27] As of 2016, some researchers consider autism to be prevalent but underrecognized in developing countries, whereas others think it is comparatively rare in non-Western countries.[28] Psychologist Tamara Daley, who has done extensive research in India, suggests a middle position in which autism is viewed "as a universal disorder which—like schizophrenia—occurs in some form in all cultures, though still susceptible to cultural influences in expression and course."[29]

Daley's suggestion makes sense. Among developing countries, India is especially well studied, and research there helps explain why accurate prevalence estimates for autism are so difficult to obtain in non-WEIRD countries.[30] For starters, arranged marriages are still the norm in India, and genetic conditions such as autism reduce marriageability in addition to increasing stigmatization and discrimination against affected individual and their mothers, who are frequently blamed for genetic conditions.[31] Further, "unlike in the US, having a diagnosis of autism may actually *reduce* the availability of services for a child in India."[32]

Similarly, anthropologist Roy Grinker of George Washington University and his colleagues explored attitudes that affect recognition of autism in children aged eighteen to thirty-six months in a low resource, Zulu-speaking community in South Africa, with results that were consistent with some of the contrasts discussed above between Western and non-Western cultures:

Parents interviewed during the study noted that the "western" diagnosis of ASD alleviated their anxieties about spiritual causation and provided a useful framework for them to understand their children's condition. However, in doing so, the diagnosis also affirms that it is the child and not the child's social network that is sick. In contrast, a parent who consults a traditional healer typically learns that it is the society not the child that is afflicted. The ancestral spirits were displeased with some event or failure to follow custom, and expressed their displeasure through the child. Appeasement and cleansing rituals are often used to placate the ancestors and reaffirm the belief that an illness emanates not from the body of the innocent child but from society.[33]

One must be cautious, however, about jumping to the conclusion that if cultural and methodological issues like social stigma and small sample sizes could be taken into account, autism would be as prevalent in developing countries as it is in WEIRD ones. A thought-provoking study by Cristiane Paula of the Federal University of São Paulo in Brazil and her colleagues suggests otherwise. The scientists carefully screened 1,470 children aged seven to twelve for autism in an urban district from southeastern Brazil and found that four of them (all boys) were autistic, which translated to a relatively low prevalence of 0.27 percent.[34] The only other reliable study in South America at that time was in Venezuela,[35] which also had a low prevalence for autism (0.17 percent) compared to the approximately 1–2 percent estimates of prevalence for the world's population of children noted above.[36] Because Paula and her colleagues controlled for the pitfalls that are usually thought to explain low prevalence of autism in non-Western countries, they concluded that its low prevalence in Brazil might be due to factors such as increased child mortality because of inadequate health care systems and higher reproductive disadvantages for survivors and their relatives that frequently characterize developing countries. If so, Darwinian natural selection combined with cultural influences may be keeping the prevalence of autism relatively low compared to other regions where autistic youngsters are not only more likely to survive and thrive but may also benefit from certain autistic traits— the Silicon Valleys of the world being an extreme example.[37] To the extent that future prognoses for autistic infants improve in developing parts of the world, one would expect its prevalence to increase.

Unfortunately, there is a scarcity of information about autism among the planet's rapidly disappearing foraging, hunting, and gathering societies. I asked Russell Greaves, an archaeologist at the University of Utah who has done ethnoarchaeological and behavioral research with Pumé foragers of the savannas of Venezuela since 1990, if he had observed children who may possibly have been autistic. In response, Greaves wrote:

> I have only encountered one girl who exhibited odd and socially inappropriate behaviors (out of a population of ~83 persons in 1990–1993, slightly more individuals now). She did not exhibit extreme withdrawal or repetitive hand or other body movements, but had some odd gaze behaviors and other anomalies. I first knew this Pumé girl at the age of ~4, and she is currently a young married woman with no surviving children (not necessarily an unusual situation among savanna Pumé who have ~35% infant mortality and additional child mortality that makes ages 0–15 a 50% mortality period). She had strange food wasting behaviors as a young child, (very unusual for this highly nutritionally stressed population), was recognized by the Pumé camp as very unusual in many behaviors, and had a very late onset of language skills. She is still quite reserved and linguistically still potentially underdeveloped.[38]

Although reports like this are anecdotal, they are interesting because lifestyles of the few remaining hunting and gathering societies are thought to reflect to some degree those from prehistory. Other examples of possible autism come from the Hadza of Tanzania, East Africa, one of the world's last remaining hunting and gathering groups.[39] Anthropologist Colette Berbesque (fig. 6.1) of the University of Roehampton in London observed two possible cases of autism among the Hadza, in one boy and one adult man (not close relatives). According to Berbesque,

> Both are intelligent, but socially awkward and do not make eye contact. The adult has yet to have a real marriage. He claims to have been married once, but says that when he went to announce it to his parents (in another camp) she was gone when he returned. If he is telling the truth, the marriage might have only lasted days. He has named his ex, but I've never found her (so I can't confirm his story of a very short marriage).

Figure 6.1. Evolutionary anthropologist Colette Berbesque in Tanzania with Hadza children. To date, little or no research has been carried out on autism in hunting and gathering societies. Hopefully, this will change before such societies disappear. Courtesy of C. Berbesque.

He is the only man that I know of that hasn't married by his age (mid-late thirties), despite being smart, reasonable looking, and generally very nice. I have attributed it to emotional intelligence and social clumsiness, because in every other respect he would be a good catch. I hope the boy is just shy, but his sort of shyness is not common among the Hadza.[40]

What would we expect autistic children in hunting and gathering societies to be like? Rather than systematically arranging toys or books like the children shown in figure 5.1 in the previous chapter, perhaps autistic children in traditional societies would be fascinated with natural objects like sticks, rocks, and plants or, similar to a boy from Croatia described below, bugs. As they grow up, these individuals might become experts on birds, other animals, or animal tracks. Hopefully, cultural anthropologists will begin to explore the prevalence and nature of autism in hunting and gathering cultures before these cultures completely disappear.

But what about the global prevalence of Aspies? Although most of the cross-cultural literature on autism does not specifically address AS, there are a few reports that suggest it occurs globally, at least to some extent. For example, a case of AS was recently described for a seventeen-year-old male in the low-income country of Croatia.[41] As a child, he showed the classic signs of AS, including excessive isolated interests—he played with ants and other bugs and tore books into strips of paper. After he entered school, he had no friends, was bullied, avoided eye contact, and spoke in a monotone voice. As of 2015 (when the report was published), he was about to graduate from high school with excellent grades. The young man was not diagnosed with AS until he was an adolescent, which the authors attributed to "a limited capacity for early identification and intervention of AS in Croatia which is a low-income country."[42]

Another interesting example of a likely Aspie (although she, apparently, was not diagnosed specifically with AS) comes from India: "Bilingualism among verbal autistic children was commonplace. One teenage participant was fluent in five languages: she heard and expressed herself the most in Bengali, followed by Malayam, then English, Hindi and Tamil, and was able to write in Bengali, English and Hindi with correct grammar. This teen automatically spoke the language of the person she was encountering, which suggests how bilingualism might be used as part of a theory of mind paradigm."[43] In another example from India, AS has been reported anecdotally for the technological community in Bangalore, India.[44]

Further evidence that AS is a global phenomenon comes from the first clinical study to examine individuals with AS from a middle-income, developing country, namely, Turkey.[45] This study focused on the types of special interests of sixty-one children and adolescents (fifty-six boys and five girls) who were diagnosed with AS using strict *DSM-IV* criteria for Asperger's Disorder. Interestingly, the children's most common interests were electronic devices, computers, and technical subjects, and their second most common areas of concentration included geography, history, and science. Other interests that were occasionally reported were cartoon heroes, music, war planes, writing and drawing, playing Lego, and holiday plans. The authors concluded that the reasons for clinical referral as well as the special interests of Aspies in Turkey were very similar to those for AS in "developed western countries, indicating the universality of symptoms."[46]

Should Western Diagnostic Criteria for Autism Be Imposed Globally?

Some scholars question whether it is appropriate for developed countries to use "deficit" medical models to diagnose autism on a global scale (for example, by applying the criteria listed in *DSM-5*), because doing so is premised on the view that autism is a mental disorder. Although details of the onset and symptoms of autism are not yet well understood across cultures, recent research suggests that the point at which individuals are judged to be abnormal varies enormously. As noted, for example, the highest-known prevalence of autism in the world (2.64 percent) has been estimated in South Korea using diagnostic criteria developed in North America and Europe.[47] However, more than half of the South Korean children assessed as autistic attended mainstream schools and had no previous history of educational or clinical problems. As psychologists Courtenay Frazier Norbury of the University of London and Alison Sparks of Amherst College ask, "If these children are succeeding in mainstream schools and are not causing concern, should we diagnose them with a disorder?"[48] An alternative, but not necessarily mutually exclusive interpretation, is that diagnoses of autism are low in South Korea, even though its prevalence is high, because it is viewed as "a hereditary disorder that impugns the genealogy and complicates marriage opportunities for extended family members."[49]

Norbury and Sparks raise other important questions about using Western criteria to assess neurodevelopment in children in non-Western cultures. For example, although it is well known that autistic individuals in North America and Europe avoid eye contact, direct eye contact in children from traditional Asian cultures is considered rude and arrogant. This shows that routinely making eye contact is not universal among children and that Asian youngsters' failure to do so should not be used to screen them for autism.[50]

When the culturally appropriate Mandarin Childhood Autism Spectrum Test was recently used to screen primary school children, it measured a prevalence of just over 1 percent for mainland China, which was much higher than earlier estimates and more in keeping with developed countries.[51] Because autism is socially stigmatized in China, autistic children might be prevented from being accepted by mainstream schools. The authors of this study suggested that the actual prevalence of autism was likely to be even higher than their estimate because academic achievement is highly valued by Chinese parents who answered the questions used to screen their children.

Another example of possible cultural differences in the manifestation of autism comes from reports that stereotypical behaviors such as hand flapping and rocking, which are common in autistic children in Western countries, may be uncommon among autistic youngsters in parts of Africa.[52] Recall that such "stimming" may, in part, be a way that autistic children in Western countries fulfill the deeply engrained need for contact comfort that emerged prehistorically when infants were periodically separated from their mothers. One wonders if these remarkable reports (if borne out) may be related to the fact that, compared to mothers in Western cultures, women in many African societies habitually keep infants attached to their bodies with baby slings.[53]

Despite the shortage of data from small-scale societies, the global evidence reviewed in this chapter suggests that autism is, indeed, a universal human condition. Nonetheless, researchers and clinicians should be cautious about using Western criteria to identify autism in populations where relevant behaviors are constrained by cultural norms like wearing babies in slings, or by taboos such as prohibitions against direct eye contact and finger pointing in some traditional Chinese cultures. As Norbury and Sparks put it,

> Should psychologists persist with a very Western medical model of disability, in which the problems faced by individuals are part and parcel of their disease, or should a more social model of disability be adopted, which focuses on society's ability to adapt to the variable language and communication needs of the population as a whole? In recent years, the neurodiversity movement has been influential in bringing these issues to public attention, advocating that ASD in particular is natural human variation and should be viewed as a separate minority culture ... as opposed to a disorder.[54]

Neurodiversity, Computers, and the Internet

In 1999, a woman with AS from Australia named Judy Singer made a clarion call that ignited the neurodiversity movement:

While autism is associated in the public mind with images of rocking, emotionally cut-off, intellectually impaired children and "Rainman"-like savants, a range of people who are not intellectually impaired, and may even be intellectually outstanding, are recognizing themselves as being "somewhere" on a continuum between "normality" and classical autism. . . . For most people with AS, . . . the autistic spectrum is above all a hypersensitivity to sensory stimuli, which necessitates withdrawal from a world of overwhelming sensation. . . . For me, the key significance of the "autistic spectrum" lies in its call for and anticipation of a politics of neurological diversity, or neurodiversity. The "neurologically different" represent a new addition to the familiar political categories of class/gender/race and will augment the insights of the social model of disability.[55]

Singer, whose mother and daughter also had AS, pointed out that communicating via the Internet was appealing for autistic individuals because it removed constraints that occur during face-to-face conversation, such as the rapid processing demanded by speech and being bombarded with eye contact and body language. Others have since emphasized that communication via computers permits a more comfortable, slower processing speed. Further, "computers are often particularly well suited for those on the autistic spectrum as they provide interactive consistency. A computer has the same response for a given input, so there [are] no body language or tone of voice messages that need to be decoded."[56] This suggests that a major benefit of using computers for social communication is that it does not require the rapid synthesizing of right-hemisphere tone of voice with left-hemisphere symbolic language that is entailed in face-to-face vocal communication between neurotypical individuals. In other words, autists can go at their own pace without feeling socially awkward.

Singer also noted the popular perception that the Internet and cyber culture were developed primarily by "nerds" and "geeks," and observed that such individuals fit perfectly on the autism spectrum. In a prediction that turned out to be prescient,[57] she further stated, "Perhaps it is not too fanciful to suggest that we are entering an era of co-evolution with machines that opens up a new ecological niche for people 'on the spectrum,' allowing them/us to flourish and come out with pride."[58] Indeed, since Singer's seminal paper, autists have developed an enormous presence on the Internet, which enables not only social communication (dispelling stereotypes about social aloofness) but also the emergence of a culture of unified positive identity and advocacy.[59] As discussed in the next chapter, Singer's prediction about autism, evolution and machines was prophetic in more ways than one.

Needless to say, the assertion that atypical neurological development is simply a part of natural human variation contradicts conventional concepts of normality and abnormality, which reflect the degree to which individual traits and behaviors coincide with, or deviate from, average or common traits and behaviors. As touched upon in chapter 1, although some find the concept of neurodiversity problematic when applied to low-functioning autism, ethicists Pier Jaarsma and Stellan Welin of Linköping University in Sweden make a strong argument that it is reasonable to apply it to HFA and AS.[60] They note that *DSM-IV* and *DSM-5* represent a medical model in which individuals are viewed as people with disabilities that should be fixed, cured and corrected. Moreover, to "subsume Asperger's Disorder into Autistic disorder in DSM-V [*DSM-5*] is a wrong way to go. To be put in the same category together with low-functioning autists may be regarded by some of the persons with Asperger's as an even worse stigmatization."[61] In fact, "some of them are being harmed by it, because of the disrespect the diagnosis displays for their natural way of being."[62]

A convincing and poignant case has also been made that AS is neither a disorder nor an undesirable condition *per se*, but instead represents a condition with a particular vulnerability, similar to that suffered by homosexuals before 1973, when the American Psychiatric Association declared that homosexuality was no longer a psychiatric disorder, and before Western society in general became less homophobic.[63] The evolutionary perspective and comparative information from both industrialized and developing cultures presented in this book lend support to the neurodiversity movement's views, at least for HFA and AS.

But there is more. As this chapter shows, the social stigmatization of autism that exists in many, if not all, cultures can lead to inaccurate diagnoses, reduced access to educational resources and social services, and underestimates of the prevalence of autism. There is also another *huge* negative spinoff, at least for children in the United States and Britain—namely, that autistic children are all too often bullied mercilessly by their peers. Aspies desire and seek social contact, even though they are keenly aware of their social awkwardness, which makes them especially prone to experiencing meltdowns and depression in response to bullying. Below, Eve minces no words about what she thinks of bullies.

EVE: ABOUT BULLIES

Other Aspies who have written books talk about school in the same way. They say that they stayed out of the way of other kids in a corner somewhere staring at the sky. They also say that they knew they were different from other kids in some nameless but all-pervasive way. I will not say any of this because it is not exactly true of me. I wasn't in a corner of the playground away from everyone else. I was on the swing set taking myself higher and higher in the stories I told to only myself. I didn't exactly know I was different from anyone else either. I just knew that for some reason a group of three or four other kids decided I was a good target for their narcissistic bullying and teasing. Of course, I didn't really understand why they said what they said to me. I just knew that it hurt a little and that I told my mom and possibly the teacher as well. I have no compunctions about being called a tattletale even now. I would rather be called that than have some tyrannical bully ruling my life or my emotions. It just wouldn't be a comfortable state of affairs.

I think my emotional experiences during childhood were richer than those of neurotypical people because my emotions seemed purer in strength than theirs. By which I mean I didn't temper my emotions with control like other children did. I tended to come out and show whether I was angry, upset, or happy. It all showed on my face somewhat. Much to the point that growing up I seemed to hear people ask me a lot, "Eve, what is wrong? Why are you upset?" or something to that effect. Normally I would answer that nothing was wrong and I wasn't upset. If I were truly upset about anything, I would usually end up having some sort of meltdown. More specifically I would start crying or on occasion when I really got annoyed I would scream like, as my father put it, "a banshee." I really don't have all that much control over the volume of my voice to the point that even now I am still told to talk in an indoor voice, most often by my younger sister Helen, who seems to view it as her job to make sure I actually act civilized.

The first school I attended was a pre-kindergarten Montessori school in America, which, from what I remember, was quite fun despite the repeated attempts from bullies to make me feel as though I had grown three noses at once or something. Some of my behavior at school might have seemed a tad outlandish or odd to the other children who it seems were all perfectly normal, at least in their mental capacities. They seemed to get some enjoyment out of calling me odd names that were to do with some of my behavior at school or they just enjoyed mimicking my stutter. I have to say that I didn't enjoy either of those things. Not that I paid much attention to what they were saying unless they were right up to my face. They only seemed to enjoy teasing me when in a group.

In my opinion bullies like that are the worst kinds of people to associate with. They ought to be ignored and ostracized by those around them, if only to give them a taste of their own bitter medicine and make them consider what better use of their time they could make than making other people around them miserable. Though I guess part of what made me such an attractive target for the bullies was that no matter what I wasn't really going to break the habits that I already had. If anything, some of what those hooligans said to me caused me to develop nervous tics or habits that I have now. These include a gradual increase in volume of my voice when I am nervous, as well as odd laughter and a tendency toward some repetitive throat clearing or phraseology. At the time, I also tended to start daydreaming or talking out loud to myself when I was nervous or just bored during class. In these daydreams I would turn my hands into two twin girls called Dina and Dana and I would tell stories about them using the plots of cartoons I had seen on television as inspiration. I guess you could call it a slightly different version of what most children use finger puppets for.

As I said before, when the sisters and I moved to Bristol, Dad decided to find a school for kids like me. The key difference between my schooling in America and my schooling in England is that in America my schooling was all mainstreamed with the general population of the school, whereas in England I was in a specialized unit for kids with Asperger syndrome and autism, which was attached to a mainstream school. This allowed me to take exams in subjects that interested me and also to learn social skills with people who actually understood me. Being in that school allowed me to come out of the shell I was sort of shut away in playing hermit. For the first time,

I had people around me who were like me, and most of them were actually nice and had manners. The school had a somewhat tolerant policy, but bullying was not permitted as much as at other schools I had attended. People who were caught doing that got school suspension. In other words, the day after they did anything that resembled bullying they were kept in one room doing work for the entire day, and in my opinion they totally deserved it.

There was one boy who got so mad so often that I was surprised he wasn't constantly getting in-school suspension for his absolutely detestable temper and behavior. I thought the only thing that interested bully boy was football, football, and yet more football. Oh, yes, he also seemed to enjoy either frightening me or hurting my feelings by calling me a variety of really hurtful names depicting my physical attributes and my background. Some of these names included "baboon," or "Yankee" because I spent over ten years living in America and had an accent from it. He is lucky I didn't decide to retaliate in kind and only told the teachers what he was doing. If I had a mean bone in my body I would have called him the "British idiot," or "skeleton boy," though I think he might have liked being called that last one.

The upshot of this was that I spent a lot of time with adults. Especially in America, when my Grandma would throw one of her amazingly good parties to which were usually invited a remarkable number of her work colleagues. I would spend a lot of time at these parties talking to the adult guests instead of any of the guests who were my own age, mostly because people my own age usually seemed to be so immature or they just didn't seem to think that I was all that good of a companion. Anyway, I preferred the company of the adults to that of the children.

Chapter 7

Asperger's Garden
What Does the Future Hold?

We can see in the autistic person, far more clearly than
with any normal child, a predestination.

—HANS ASPERGER[1]

Full Cycle—Back to Evo-Devo

Although one can either think of autism as embodying one large spectrum[2] or recognize that it has various subtypes, these approaches need not be mutually exclusive. Both are useful when thinking about the evolution of autism, which may affect up to 2+ percent of some of the world's populations—and still counting. By looking across the entire spectrum, for example, it is clear that autism is intimately bound up with language or its impairment. Roughly speaking, an estimated 50 percent of autists remain mute,[3] approximately 40 percent are delayed in acquiring speech, and the remaining 10 percent develop fully grammatical language in a timely manner, although it is often expressed with an atypical tone of voice and may be associated with enhanced vocabulary and reading skills. Partly because autism is so entangled with language, it is probable that its neurological and genetic underpinnings emerged on the coattails of prolonged linguistic evolution and its more recent cognitive spinoffs.[4] As we have seen, this idea fits well with comparisons of DNA in contemporary people and their ancient cousins, the Denisovans and Neandertals, including findings for chromosomes 16 and 7 (among others). Thus parts of *Homo sapiens'* genome "contain genes linked to language, brain development, and autism, [which] are critical to a modern human's identity and reproductive fitness: Archaic gene variants can't be tolerated here."[5]

It therefore seems likely that autism first appeared in our prehistoric relatives after *Homo sapiens* diverged from the evolutionary branches that led to other, now extinct, ancient humans, including the Neandertals and Deniso-vans.[6] Nonetheless, the relatively recent appearance of autism during homi-nin evolution does not obviate the fact that evo-devo precursors from a much deeper past paved the way, not only for its appearance but also for our species' current tolerance of a large number of comparatively rare genetic changes that are associated with a risk for developing it. As science journalist Steve Silberman so eloquently put it, "Whatever autism is, it is not a unique prod-uct of modern civilization. It is a strange gift from our deep past, passed down through millions of years of evolution."[7] Further, when it comes to individ-uals with AS and HFA, this strange gift may have a surprisingly large impact on the future evolution of *Homo sapiens*.

Neurocultural Evolution

Evolutionary biologists agree that humans are continuing to evolve. Some of the clearest evidence for this comes from genetic changes in populations in response to infectious diseases such as malaria or HIV. What has not been widely recognized, however, is that humans have experienced enormous cul-tural changes that were (and are) intertwined with recent cognitive, genetic, and neurological evolution. Heritable biological changes that not only turn genes on and off (via epigenetic mechanisms)[8] but may also accelerate the pace of evolution[9] have likely played a role in the recent, relatively rapid evolution of humans. Because some of our species' most startling neurolog-ical adaptations have been driven by cultural innovations, I think of them as resulting from neurocultural evolution in much the same way that the changes that occurred in the frequencies of hemoglobin genes in African populations after they took up farming are attributed to biocultural evolu-tion. As discussed below, the processes responsible for recent neurocultural evolution in humans are, if anything, speeding up, and this should bode well for the future of high-functioning autists.

Reading: A Prime Example of Recent Neurocultural Evolution

Without writing and the ability to decipher (read) it, our ancestors would not have been able to externally archive, recall, and process the vast amount of information that facilitated the emergence of *Homo sapiens* as the planet's dominant species. It is remarkable, then, that as far as we know, the world's first full-fledged writing system did not appear until a mere fifty-five hundred years ago. That debut featured a script for the Sumerian language of ancient Mesopotamia (modern Iraq), although it had earlier precursors in the form of abstract tokens used to facilitate trade, famously discovered by archaeologist Denise Schmandt-Besserat.[10] Despite the fact that illiteracy is considered a serious problem, as of 2012 the global literacy rate for adults was estimated to be 84.3 percent, and that for youth was even higher at 89.4 percent.[11] I find these figures astonishing, given that reading and writing were invented so recently.

Some have suggested that literacy did not actually evolve but, instead, simply arose in response to cultural influences. The evidence, however, suggests otherwise. For starters, a good deal of research has shown that dyslexia has complex genetic substrates and is highly heritable.[12] Studies that compared reading acquisition in typically developed identical and fraternal twins confirm that, although reading is influenced by environmental factors, genetic influences are highly significant and stable as "children move from learning how to read to using reading for learning. . . . Thus, although reading is clearly a learned skill and the environment remains important for reading development, individual differences in reading comprehension appear to be also influenced by a core of genetic stability that persists through the developmental course of reading."[13]

It's not just reading that is highly heritable; so is much of the neurological wiring that it depends upon. As French cognitive neuroscientist Philippe Pinel and his colleagues put it, "Although it may seem paradoxical that a cultural acquisition such as reading depends on a heritable brain area, many recent studies have, in fact, found that reading acquisition is supported by a tightly constrained preexisting architecture for spoken language processing and visual recognition."[14] In other words, the genetic and neurological underpinnings of reading piggybacked on ancient substrates that initially developed with the invention of speech.

How the brain processes reading is fairly well understood.[15] Written words or characters (such as those in Chinese) are perceived with the eyes, which send information to the visual cortex at the back of the brain. From there, the script is sequentially decoded, mostly on the left side of the brain, in a progressive manner as relevant information flows forward in the brain. In English, for example, the back of the brain recognizes individually meaningless visual fragments that are components of letters (such as lines with certain orienta-tions). As this information is processed and sent forward, visual fragments are assembled into letters, then the letters are interpreted as words, the individual words are associated with concepts, and, finally, strings of words are interpreted semantically. Although there are subtle differences for languages that use char-acters instead of alphabets,[16] this is the basic game plan for reading any script in the world. Because all writing is the visual representation of speech, it is not surprising that, universally, "the reading network is deeply constrained by the organization of the brain network underlying speech."[17]

It would be a mistake, however, to think that reading relies exclusively on one simple forward-flowing neurological pathway. Reading depends on input from various neurological networks, such as those involved in identifying speech sounds, rapidly naming familiar visual symbols, and holding infor-mation in short-term memory, to name just a few.[18] In particular, two major pathways in the left hemisphere facilitate reading.[19] The so-called ventral pathway is implicated in processing whole words visually, and involves parts of the occipital and temporal lobes including an important region called the "visual word form area" (VWFA), which is part of the left fusiform gyrus.[20] Another route that is located higher up in the left hemisphere (the dorsal pathway) converts written symbols to sound, which can be spoken silently or aloud and then processed for meaning. This pathway is obviously important for sounding words out. Not surprisingly, successful reading involves flexible shifting between these left-hemisphere pathways.[21]

One can't help but wonder what the above networks were doing before the invention of writing. Although it is speculative, the VWFA in the left fusi-form gyrus may have contributed to processing faces, which is done today mostly by a region in the right hemisphere called the "fusiform face area" (FFA). Interestingly, the VWFA and FFA in the left and right hemispheres, respectively, occupy similar positions in the fusiform gyri, and both process visual stimuli "consistent with the need for high spatial accuracy in decipher-ing the fine details of these stimuli."[22] Further, as individuals learn to read, the VWFA in their left hemispheres matures and the processing of faces comes to depend more on the right than left FFA.[23] All this is consistent with the suggestion that, as reading evolved during the last fifty-five hundred years, it hitchhiked on left-hemisphere language networks and displaced much (but not all) of the processing of faces that goes on in the left FFA to the right hemisphere. If so, reading has contributed to the ongoing trend for increased brain lateralization in humans. The same may be said, by the way, for the relatively recent invention of arithmetic.[24]

The reader will recall that, although Aspies are generally not good at interpreting facial expressions, they may be obsessive readers who taught themselves to sound out words at a very young age. Despite the fact that clinicians tend to view such precocious reading as pathological, some researchers have argued persuasively that this trait is, in fact, a "superability."[25] It is not surprising, then, that the ventral pathway for visually decoding words and the dorsal pathway for sounding them out are equally activated in the brains of adult Aspies when they read, whereas the ventral pathway is much more active than the dorsal one when TD adults read.[26] As we have seen, reading is a comparatively recent neurocultural innovation that was associated not only with stunning cultural changes, but also with neurological rewiring. We don't know if Aspies were among the Babylonian bookkeepers who first invented writing, but we do know that when it comes to the neurological substrates for reading, they did well in the evolutionary sweepstakes.

Computers: Harbingers of Future Neurocultural Evolution

During this past spring break at Florida State University, where I teach, I decided to visit my sister who lives near Orlando, so I hopped on a bus in Tallahassee for the five-hour trip. I was surrounded by FSU students, who also were going on break. All around me, they sat silently huddled with pillows and blankets, earbuds in, listening to and/or watching media on portable smart devices or reading and texting on them. No one spoke, looked out the window, or did much of anything other than move their fingers to text—for the entire trip. What, I wondered, is happening to their brains? This turns out to be a controversial question. Some believe that use of the Internet is unlikely to be associated with cognitive and neurological changes,[27] but others demur.[28]

Research supports the latter view. Although the Internet only became available to the public within the last twenty-five years, studies show that children who have grown up using it differ cognitively from (frequently older) individuals who adopted it later in life. Use of the Internet has thus been associated with a shift "towards a shallow mode of learning characterized by quick scanning, reduced contemplation and memory consolidation," and greater distractibility.[29] Significantly, some believe the recent shift to fast-paced digital media (which may include text, audio, video, and graphics) is interfering with the acquisition of the "deep reading" that facilitates comprehension, reasoning, and critical reflection.[30] On the other hand, use of the Internet also appears to enhance users' multitasking skills,[31] and action video games, in particular, have been shown to improve gamers' visual attention compared to nongamers.[32] So far, few studies have investigated the relationship between use of digital media and altered brain connectivity. Nonetheless, preliminary evidence shows that regular use of the Internet is, indeed, associated with neurological reorganization, which hitchhikes and expands on reading networks, particularly in the left hemisphere.[33]

And make no mistake, computers and digital media are set to impact human society at least as much as reading has. As of 2016, the world's leading online game, *League of Legends*, attracted twenty-seven million players each day, despite the fact that it had a steep learning curve.[34] By 2020, an estimated 80 percent of the world's adults are likely to own smartphones, and driverless cars, drone deliveries, incredibly sophisticated robots, and increasingly improved navigational systems are just around the corner.[35] Will human brains evolve in response to these technologies? If the past is the best predictor of the future, of course they will. Not all the changes are likely be positive, however. "If we do not cherish them," one author cautions, "our natural navigation abilities will deteriorate as we rely ever more on smart devices."[36]

The fact that the Internet and digital media have had such a tremendous impact in less than a quarter of a century suggests that humanity may be beginning an episode of neurocultural evolution similar to the one that started about fifty-five hundred years ago and sparked the eventual global emergence of reading—only this time the impetus will be the availability of computers rather than books (or scrolls). Further, one has the sense that neurocultural evolution may be speeding up as it builds on itself: first came spoken communication, probably hundreds of thousands if not millions of years ago; writing arose less than six thousand years ago; and public Internet communication made its debut just a little over two decades ago. What could be next—direct mind-to-mind communication?[37]

How is such an escalation in neurocultural evolution possible? Unfortunately, the mechanisms whereby widely adopted novel behaviors such as reading rewire "plastic" human brains and become genetically encoded (recall that reading and the neurological wiring that support it are highly heritable) are not well understood. As noted, recent evidence suggests that these mechanisms likely involve the transmission of epigenetically tweaked (but not basically altered) genes across generations. This idea is consistent with the fact that epigenetic modifications mediate the brain's plasticity, learning, and memory and that such tweaks are passed along to human offspring more frequently than previously thought.[38] Once invented, it seems plausible that reading eventually became highly heritable because it was associated with successful survival and reproduction. As noted, it is also possible that epigenetic inheritance may have accelerated such evolution.[39] In other words, rapid neurocultural evolution may well entail synergy between epigenetic inheritance and natural selection.

And where are Aspies in all of this? As science journalist Steve Silberman notes, "It's a familiar joke in the industry that many of the hardcore programmers in IT [information technology] strongholds like Intel, Adobe, and Silicon Graphics—coming to work early, leaving late . . . —are residing somewhere in Asperger's domain."[40] Further, it seems likely that incentives for Aspie employment in tech-related jobs will increase.[41] Thus, according to economist Hays Golden, "given how many high achievers have Asperger's, I believe that there is a strong argument that the costs of having Asperger's are below the earning potential of someone with Asperger's,"[42] and that "developments such as computers . . . have likely increased the returns [wages] to systemizing."[43] It appears that the prevalence of Aspies (and people with HFA) is, indeed, rising in at least technological enclaves, as suggested by psychologist Simon Baron-Cohen.[44] If one puts all this together with the above information about the neurocultural evolution of reading, it seems safe to predict that AS will eventually become much more prevalent and socially acceptable than it currently is. In fact, if we time traveled fifty-five hundred years into the future, who's to say that we wouldn't find that AS had become the new normal?

Back to the Present

Hans Asperger would not have been surprised by this book because, unlike many contemporary clinicians and researchers, he emphasized the remarkable intelligence of his patients and went so far as to speculate that the self-absorption and "increased personal distance" of children with AS were prerequisites for developing the original perspectives that sparked their creativity.[45] Today, it is accepted that a certain amount of social cluelessness counterbalances Aspies' focused attention on their inner worlds. Fortunately, people with AS can readily learn many of the social norms they fail to develop spontaneously. If you met Eve, for example, she would offer you her hand, look you in the eye, and say something like "nice to meet you"—a script that was given her by teachers who specialize in helping autistic children learn basic living skills.

The average age at which AS is diagnosed is considerably later than that for other forms of autism,[46] but this is changing as new methods are discovered for detecting autism in very young infants.[47] Earlier recognition, of course, leads to earlier intervention. Significantly, some clinicians have become attuned to the evolutionary framework within which autism arose[48] and are beginning to recognize the use of motherese as a valuable therapeutic tool for dealing with youngsters who are likely to develop autism.[49]

Despite the increasing awareness of autism, there is an urgent need for more public education about it and for more clinicians and teachers to help treat and assist the autistic population. Aspie and HFA children are all too frequently the targets of brutal teasing and bullying, and they may continue to be stigmatized as adults, especially in industrialized societies such as the United States and United Kingdom. The disparagement that many people with AS and HFA experience is extremely dehumanizing, as described by leaders of the neurodiversity movement. No wonder these children grow up preferring their own company, and no wonder they are at risk for depression. Equally worrying, researchers have found that children with AS have more difficulty controlling their social-emotional responses than other autistic children, which may be "related to the tantrums, rage, and meltdowns often exhibited by children and youth with AS."[50] Bullying can cause meltdowns, of course, and these have the potential for being very costly, not only for those directly involved but also for their loved ones and, sometimes, society at large.

On a more positive note, the neurodiversity movement is having a positive impact on how autism, including AS, is perceived, and unprecedented numbers of autists at the higher end of the spectrum are networking socially via the worldwide web. Equally encouraging, exciting new gene-editing techniques are being used to create mutations in animals that mimic those associated with autism in humans, with an eye toward eventually discovering pharmacological treatments for autists.[51] The neurodevelopmental trajectory of autism continues to be explored with increasingly refined brain imaging techniques, and research specifically on AS continues full throttle despite the removal of Asperger's Disorder from *DSM-5*. Furthermore, each new discovery about the cultural, behavioral, neurological and genetic underpinnings of AS adds to the likelihood that "Asperger's Disorder will be back" as an official diagnosis in the next edition of the *Manual*.[52]

Eve and I hope that this book will contribute to an increased awareness that Aspies are endowed with advanced cognitive abilities that were pivotal for making us human. As we have seen, microevolutionary and biocultural factors continue to influence the prevalence of autism in both developed and underdeveloped countries. Given the direction that human neurocultural evolution is headed, it seems likely that autism will continue to increase globally, and that individuals with AS and HFA will play an increasingly important role in the evolution of our species.

EVE: HELPING YOUR ASPIE CHILD HAVE A BETTER FUTURE

I personally think that I had a good and fun childhood with only a few things that needed improving, but I have some specific suggestions about what parents of Aspies can do to avoid making mistakes. For example, find a school with an inclusion or mainstreaming program so that your child can have some semicontrolled contact with other kids who don't have AS. If you can't do this, try to find a school specifically for kids with AS because I can tell you that it really helped to be around kids who sympathized with what I was going through, especially after attending at least three different mainstream schools. I would also advise parents to find a school that offers a foreign language or foreign cultures program, because cultivating a curiosity about other cultures and other languages is good for Aspies. In my case, the Montessori school I attended as a kid taught us how to speak some basic French, and I enjoyed that immensely. The middle school I attended taught both French and Spanish, and I had a bunch of games and CD-ROMS as a kid that taught me a mixture of those two languages. Currently I am teaching myself how to speak Japanese by watching anime with English subtitles.

It is also a good idea for parents of children with AS to encourage them to use their obsessions or primary interests in whatever career they choose to follow in life. To this end I would also advise parents to help their children figure out what they want to do by at least middle school so they can pick their school subjects accordingly. In my case, I could have used this advice about seven years ago.

Parents should also encourage their Aspie children to do a sporting activity. Doing some sort of team sport is a good and extremely controlled way to socialize with other people in that there are rules for one's conduct during sporting events and there are certain expected behaviours when one is socialising with teammates or fellow athletes. Getting started on doing some sort of sport or regular exercise will help your Aspie child learn about his or her body's physical limitations and whereabouts.

I do tae kwon do, which is a mix of karate and kick boxing that originated in Korea. It was developed for use by the military, which probably explains why the concepts of discipline and respect are valued so much and why rank can be so important in this martial art. That is to say, students are ranked by belt colour and how long they've studied the martial art. The belt colours in order from lowest to highest are white, which signifies one's status as a beginner, yellow, green, blue, red, and finally black. To get each of these belts you have to attend a grading where you are tested on your knowledge of the classwork you have done for that level, which for white belts is usually a simple pattern. You are also tested on basic kicks and punches as well as verbal questions after the practical or physical part of the grading. For higher grades, the grading includes focus kicks, which showcase your control, and possibly sparring. Of course, you don't go from one belt colour to another right away, even when passing a grading, because there is an intermediate rank or grade in between each belt colour called a tag. The order for the tags is the same as the order for the belt rankings up to black.

I really enjoy this martial art, mostly because it helps relieve pent-up emotions and stress, and it allows me to deal with a few odd urges for violence. I also like it because I get to learn some words of a new language. In tae kwon do class we learn the Korean words for different target areas of the body, the body parts used for the strikes on our opponents, and the meanings of different words specific to our martial art. These words, and the meanings of the different patterns, as well as other aspects of tae kwon do,

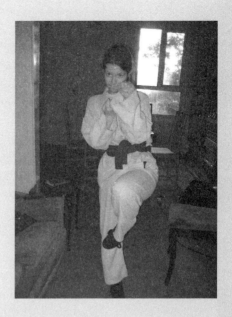

Figure 7.1. Eve doing tae kwon do.

are part of the questions asked at the gradings, along with some trivia about the martial art, such as who created it, what it was created for, and the five tenets every student of tae kwon do must adhere to, among other pieces of knowledge anyone of a particular level of skill in tae kwon do is supposed to have. I am also fond of this martial art because it allows me to get out and meet people, which helps to expand my circle of acquaintances.

As for the effects of tae kwon do physically and mentally, there are several of each. For the physical effects, I have to say that I am getting fitter and developing some muscles. I am also becoming more aware of my own body, which is to say I now know where I am situating myself in relation to others some of the time. The martial art is also helping me improve my balance. The mental effects are that I am becoming more confident, less prone to stress-related outbursts, and much less prone to letting my emotions rule me. The other martial arts I would like to study include kung fu, tai chi, judo, and jujitsu.

My final piece of advice for Aspies' parents is that you shouldn't try to handle your child's emotional development by yourself. That's a fast track to arguments and strife within the family. Hire a good therapist or counsellor to help you come up with a feasible plan for teaching your child how to keep his or her temper in check. This can be especially useful when dealing with bullies, who seem to always pick up on the fact that someone's behaviour or emotions aren't as developed as they should be. In my case, I wish my parents had done this. Call me childish, but if I'm upset I'd rather scream, shout, and cry than hold everything in until violence or breaking something is the only way to let the frustration and anger out.

Gran asked me to make a list of what parents of children with Asperger syndrome need to know, or do, to help their children have a good future. This is what I wrote:

1. Aspies may not always understand what is being talked about around them.
2. They may sometimes dominate conversations.
3. They may have specific coping strategies or habits that seem disconcerting or strange to others. Let them do these things.
4. Some tastes or textures in food will cause Aspies to say they strongly dislike them.
5. Some hypersensitivity may remain from when they were younger.
6. By the time they are teenagers, Aspies may have come out of their shell a little socially, but this doesn't mean they're ready for serious relationships with the opposite sex.
7. There may still be problems with temper control.
8. Be gentle about shifting them out of a high-concentration state of mind. They might not take being interrupted well.
9. Be careful with your language. Aspies might take what you say seriously.
10. Try to limit teasing, because even if it's affectionate, they may not like it.
11. Try to encourage their scientific interests.
12. Try to forewarn or buffer them against severe changes in routine.
13. Try to give Aspies their own space.
14. Try to involve them in the real world instead of letting them languish in their minds.
15. Push them to learn some life skills so that they can become independent.
16. Give Aspies as much love as possible.

Preface

1. American Psychiatric Association, *DSM-5*.
2. Linnaeus, "1735. Systema naturae."
3. Zeliadt, "Where the vocabulary of autism is failing."
4. Des Roches Rosa, "Listen to families."
5. Asperger, "Problems of infantile autism," 47.
6. Ruthsatz and Stephens, *Prodigy's cousin*, 7.
7. Singer, "Why can't you be normal," 63–64.
8. Silberman, *Neurotribes*, 441.
9. Singer, "Why can't you be normal," 63–64.
10. Willey, *Pretending to be normal*.
11. Willey, *Asperger syndrome in adolescence*, 12.

Introduction

1. Asperger, "Autistic psychopathy in childhood."
2. Kanner, "Autistic disturbances of affective contact." Although it is generally believed that the two men knew nothing of each other's work, new information suggests that Kanner knew about but failed to credit Asperger's research (Ruthsatz and Stephens, *Prodigy's cousin*; Silberman, *Neurotribes*, 183–86). As detailed by Silberman and contrary to popular opinion, Kanner did not name "early infantile autism" until 1944 (Kanner, "Early infantile autism").
3. Asperger, "Problems of infantile autism," 48–49. Despite his emphasis on the intellectual abilities of his patients, Asperger was aware that a minority of the more than two hundred cases he studied over ten years were people "with considerable intellectual retardation in addition to autism" (Asperger, "Autistic psychopathy in childhood," 87).
4. American Psychiatric Association, *DSM-IV*. Asperger's was also included as an official diagnosis by the World Health Organization in 1992 (World Health Organization, *ICD-10 classification*). The 1994 criteria for Asperger's Disorder from *DSM-IV* are the focus here because they are central to the controversy that continues to surround AS.
5. American Psychiatric Association, *DSM-5*. As detailed in this 2013 edition of the manual, "Autism spectrum disorder encompasses disorders previously referred to as early infantile autism, childhood autism, Kanner's autism, high-functioning autism, atypical autism, pervasive developmental disorder not otherwise specified, childhood disintegrative disorder, and Asperger's disorder" (quoted from *DSM-5*; topic: Autism Spectrum Disorder, first paragraph

under "Diagnostic Features," http://dsm.psychiatryonline.org.proxy.lib.fsu. edu/doi/full/10.1176/appi.books.9780890425596.dsm01#x98808.2728600).

6.   Kaland, "Should Asperger syndrome be excluded?"
7.   Steinberg, "Asperger's history of overdiagnosis."
8.   Tanguay, "Autism in DSM-5."
9.   American Psychiatric Association, "ASD fact sheet."
10.   *DSM-IV* allows a person to fit criteria for both Asperger's and Autistic Disorder, but requires that such individuals be diagnosed with the latter. Thus *"Asperger's Disorder* can be distinguished from Autistic Disorder by the lack of delay in language development. Asperger's Disorder is not diagnosed if criteria are met for Autistic Disorder" (American Psychiatric Association, *DSM-IV*, 69). This confusion stems from the fact that, whereas language is, by definition, not delayed in Asperger's Disorder, *DSM-IV* waffles on the criterion for delayed language in Autistic Disorder by making it optional: "There *may* be delay in, or total lack of, the development of spoken language (Criterion A2a)" (American Psychiatric Association, *DSM-IV*, 66; emphasis added). Although it seems unlikely that many people would fit criteria for both disorders, this confusion would likely result in fewer people receiving a diagnosis of Asperger's than would otherwise be the case (Szatmari et al., "Autism and Asperger syndrome").
11.   Bucaille et al., "Cognitive profile in adults with Asperger syndrome."
12.   Chiang et al., "Meta-analysis," 1577.
13.   Ibid. The IQs of Aspies overlap with the average and high average HFA IQ ranges but also include some scores in the superior range (120–29); see the graph on page 1584 of the cited article. A partial overlap in the ranges of IQs of people with AS and HFA is not surprising because of *DSM-IV*'s fuzzy diagnostic criteria, discussed in note 10.
14.   Although Grandin, who earned a doctorate, has been variously described in the literature as having either HFA or AS, the former applies because, as she described, "at two and a half, I had no speech and no interest in people" (Grandin, *Thinking in Pictures*, caption for second photograph).
15.   Foley-Nicpon, Assouline, and Stinson, "Gifted students with autism and Asperger syndrome."
16.   Charlop and Haymes, "Speech and language acquisition"; Rutter, "Language disorder and infantile autism."
17.   Ghaziuddin and Mountain-Kimchi, "Defining the intellectual profile"; Tsai, "Asperger's disorder will be back"; Tsai and Ghaziuddin, "Forward into the past"; Chiang et al., "Meta-analysis."
18.   Ghaziuddin and Mountain-Kimchi, "Defining the intellectual profile"; Planche and Lemonnier, "Children with high-functioning autism and Asperger's."
19.   Planche and Lemonnier, "Children with high-functioning autism and Asperger's," 939. Unfortunately, the authors' recommendation that the criteria for

Asperger's Disorder be revised in *DSM-5* in lieu of combining it with other forms of autism within ASD fell on deaf ears, especially since they identified specific therapeutic measures that might benefit each group. Thus "an early psychomotor therapy may be beneficial to children with AS whereas an early and specific speech therapy may improve the daily adaptation of children with HFA" (Planche and Lemonnier, "Children with high-functioning autism and Asperger's," 939).

20. Tsai, "Asperger's disorder will be back."
21. Bucaille et al., "Cognitive profile in adults with Asperger syndrome."
22. Asperger, "Problems of infantile autism"; Bennett et al., "Differentiating autism and Asperger syndrome"; Szatmari et al., "Similar developmental trajectories"; Paynter and Peterson, "Language and ToM."
23. Falk, "Evolution of brain and culture"; Falk, "Baby the trendsetter."
24. Grandin, *Thinking in Pictures.*
25. Baron-Cohen, "Autism and the technical mind."

Chapter 1

1. The opening epigraph is from Asperger, "Problems of infantile autism," 1.
2. Lee and Devore, *Man the hunter.*
3. Slocum, "Woman the gatherer."
4. Trevathan and Rosenberg, *Costly and cute.*
5. Falk, "Baby the trendsetter."
6. Green et al., "Draft sequence of the Neandertal genome"; Meyer et al., "High-coverage genome"; Prufer et al., "Complete genome sequence of a Neanderthal."
7. Falk, "Evolution of brain and culture"; Falk, "Baby the trendsetter."
8. However, there is not a consensus that the split between hominins and chimps occurred about five to seven million years ago. Adrienne Zihlman estimates a more recent date of four to five million years ago based on molecular evidence, fossils, anatomy, biogeography, and ecology (Zihlman and Underwood, *Comparative ape anatomy and evolution*). On the other hand, Langergraber et al. ("Generation times in wild chimpanzees and gorillas") estimate a much earlier split for the human and chimpanzee lineages, at between seven and thirteen million years.
9. Lieberman, "Human locomotion and heat loss."
10. Marlowe, "Central place provisioning."
11. Barak et al., "Trabecular evidence."
12. Rolian, Lieberman, and Hallgrímsson, "Coevolution of human hands and feet."
13. Interestingly, humans also have slightly longer thumbs and shorter fingers than their last common ancestor with chimpanzees (Almecija, Smaers, and Jungers, "Evolution of human and ape hand"), perhaps at least partly as a

genetic byproduct of strong selection on the toes (Rolian, Lieberman, and Hallgrímsson, "Coevolution of human hands and feet").

14. Rolian, Lieberman, and Hallgrímsson, "Coevolution of human hands and feet." As noted, the evolution of human hands seems to have been at least partly genetically linked to that of the feet. Some speculate that human hand anatomy occurred with the advent of "habitual" bipedalism (i.e., rather than at the origin of bipedalism) in association with increased manipulative ability, but that it "almost certainly preceded" the regular production of stone tools (Almecija, Smaers, and Jungers, "Evolution of human and ape hand," 8). These findings are consistent with the idea that the feet initially "kicked off" the evolutionary changes associated with bipedalism, which eventually affected the entire body.

15. Futagi and Suzuki, "Neural mechanism," 81.

16. Grasping reflexes in both hands and feet are much stronger in nonhuman primate infants than in human babies, and are replaced with more voluntary grasping as their brains mature.

17. Teulier, Lee, and Ulrich, "Early gait development." The stepping reflex is elicited by holding a baby upright and touching its foot to a mat, which causes the baby to move its legs as if it were walking. This reflex does not develop fully until around the time of autonomous walking, that is, after the foot's grasping reflex has been inhibited in human babies.

18. These reflexes provide just one example of how "the continuous and cascading interactions among many subsystems, from individuals' intrinsic characteristics (e.g., genetic, neural integrity, skeletal structures) and their environment (from the cellular environment to the outside world) shape infant development and acquisition of skills" (Teulier, Lee, and Ulrich, "Early gait development," 447).

19. Plooij, *Behavioral development*.

20. It is important to note, however, that human babies aren't nearly as delayed in their non-locomotor development. In fact, chimpanzee and human infants are remarkably similar in the timing of numerous developmental stages such as helplessness at birth, distress at separation from mother, disappearance of blind rooting responses, production of social faces, and fear of strangers (Plooij, *Behavioral development*).

21. Ross, "Park or ride?" It was not just the lack of grasping feet that made it increasingly difficult for our ancestors' infants to attach themselves to their mothers, however. With the evolution of upright walking, mothers' backs no longer provided relatively horizontal surfaces on which infants could easily ride. Further, by three to four million years ago, there may also have been less hair on mothers' backs and bellies for babies to cling to, as indicated by genetic studies of lice (Reed et al., "Pair of lice lost"). Needless to say, the evolutionary loss of infants' ability to cling would have hindered their mothers' ability to use all four limbs freely while transporting infants, for example when

climbing trees in order to build sleeping nests at the end of each day (DeSilva, "Brains, birth, bipedalism").

22. Hrdy, "Of marmosets." The extent to which cooperative breeding was involved during early hominin evolution is much debated. Although mothers in some species of primates share parenting activities (such as feeding and transporting infants) with others (siblings of infants and/or adults of either or both sexes), mothers of unweaned chimpanzees are reluctant to do so, and the recognition of fatherhood may be a relatively recent evolutionary phenomenon. Such aunting behavior or "alloparenting," however, is common among modern humans.

23. Wall-Scheffler, Geiger, and Steudel-Numbers, "Infant carrying." An interesting modern legacy of prehistoric mothers' shift to actively carrying infants is that women may be seen the world over carrying, not only babies, but other heavy loads on their heads or backs.

24. Zihlman, "Women as shapers."

25. Harlow, "Nature of love."

26. Plooij, *Behavioral development*, 45–46.

27. Bard, "What is the evolutionary basis for colic?" Although chimpanzees do not cry much, young chimp infants do when raised with inadequate maternal care. However, when these infants receive cradling contact, they stop crying.

28. Bowlby, *Attachment and loss*.

29. Provine, Krosnowski, and Brocato, "Tearing."

30. Small, *Our babies, ourselves*, 156.

31. Falk, "Prelinguistic evolution"; Falk, *Finding our tongues*, 30.

32. Schmidt and Cohn, "Human facial expressions."

33. This is probably why some parents resort to taking a crying baby for a car ride or placing her in a baby carrier that is on top of an operating clothes dryer as methods for calming and putting the infant to sleep.

34. Wermke, Leising, and Stellzig-Eisenhauer, "Relation of melody complexity."

35. Falk, "Prelinguistic evolution"; Falk, *Finding our tongues*.

36. Fernald, "Human maternal vocalizations." Compared to speech directed toward more mature individuals, modern motherese is exaggerated, slower, more musical, generally higher-pitched, and uses a simple vocabulary that has special words like "boo-boo" and "doggie."

37. I hasten to add that human baby talk evolved out of neurological substrates and behaviors that have a deep evolutionary past as indicated by infant vocalizations in some nonhuman primates that progressively develop in tune with vocal feedback from mothers. This "baby talk," however, lacks the complexity of human motherese. Takahashi et al., "Marmoset monkey vocal production."

38. For details see Falk, "Prelinguistic evolution"; Falk, *Finding our tongues*.

39. Preuss, "Human brain"; Semendeferi et al., "Spatial organization of neurons"; Teffer and Semendeferi, "Human prefrontal cortex."

40. Harmand et al., "3.3-million-year-old stone tools"; McPherron et al., "Evidence for stone-tool-assisted consumption."

41. Monastersky, "Anthropocene."

42. Passingham, "Changes in the size."

43. Knickmeyer et al., "Structural MRI study of human brain development."

44. Human fetuses have larger brains than chimpanzee fetuses, but the rate of brain growth is similar in the two species until about the twenty-second week of gestation. At that point, brain growth in the human fetus begins to accelerate compared to chimpanzees and continues doing so throughout the brain spurt (Sakai et al., "Fetal brain development").

45. Passingham, "Changes in the size"; Schultz, "Relative size of the cranial capacity." Schultz provides graphs that illustrate the nested and similarly shaped brain-growth curves for many primates, including humans and various species of monkeys and apes.

46. Gilmore et al., "Regional gray matter growth"; Knickmeyer et al., "Structural MRI study of human brain development."

47. Squire et al., *Fundamental neuroscience.*

48. Hermoye et al., "Pediatric diffusion tensor imaging"; Knickmeyer et al., "Structural MRI study of human brain development."

49. Washburn, "Tools and human evolution"; Rosenberg and Trevathan, "Evolution of human birth"; Rosenberg and Trevathan, "Birth, obstetrics."

50. Karmiloff and Karmiloff-Smith, *Pathways to language.*

51. Kuhl, "New view," 11850.

52. Mesgarani et al., "Phonetic feature encoding."

53. Kuhl, "Early language acquisition."

54. Grammar is a broader concept than syntax because it includes the rules for constructing words from components, for example, by adding endings such as "ed" or "s" to words in order to indicate tense or plurality, in addition to rules for arranging words into proper phrases and sentences. By definition, youngsters begin to engage in grammatical speech as they conform to the conventional rules for their languages, usually by the time they can produce about 150–200 words (Falk, *Finding our tongues*, 104).

55. Anderson, *After phrenology*; Grodzinsky and Nelken, "Neuroscience."

56. Gotts et al., "Two distinct forms of functional lateralization."

57. Falk, *Finding our tongues.*

58. Fiser and Aslin, "Statistical learning; "Kirkham, Slemmer and Johnson, "Visual statistical learning."

59. Rueckl et al., "Universal brain signature," 15510.

60. Schore, "Right brain is dominant," 389.

61. Gilmore et al., "Regional gray matter growth." Interestingly, this article documents cerebral asymmetries in neonates that are reversed from those seen in adults.

62. Falk, "Baby the trendsetter."

63. Ozonoff et al., "Gross motor development."
64. Cederlund and Gillberg, "One hundred males."
65. Barbeau et al., "Comparing motor skills"; Weimer et al., "'Motor' impairment in Asperger syndrome."
66. Padawer, "Recovered."
67. Crespi, "Developmental heterochrony," 6.
68. Ibid., 1.
69. Grandin and Scariano, *True story*, 28.
70. Harlow, "Nature of love."
71. Kuhl et al., "Links between social and linguistic processing."
72. David Cohen, personal communication, January 5, 2015.
73. Cohen et al., "Parentese prosody," Ouss et al., "Infant's engagement and emotion"; Saint-Georges et al., "Motherese in interaction"; Tanguay, "Autism in DSM-5."
74. Lainhart, "Brain imaging research." Studies show that this trend is not as widespread as previously believed, however. Use of improved community-based comparison samples reveals that significant brain overgrowth is found in only about 15 percent of the infants who eventually develop ASD, most of them male, which is less than previously believed (Raznahan et al., "Compared to what?"; Campbell, Chang, and Chawarska, "Early generalized overgrowth"). Further, young infants who go on to develop autism also tend to exhibit overgrowth in body weight and stature (Green, Dissanayake, and Loesch, "Review of physical growth"), although this "by no means undermines the importance of the impact that accelerated brain growth in infancy may have on the development of neuronal architecture and behavior of toddlers with ASD" (Campbell, Chang, and Chawarska, "Early generalized overgrowth," 1069). It is also worth pointing out that, like advanced cognition, increased stature was naturally selected during human evolution (Joshi et al., "Directional dominance on stature").
75. Lange et al., "Longitudinal volumetric brain changes."
76. A similar trend was found by Courchesne, Campbell, and Solso, "Brain growth across the life span."
77. Courchesne et al., "Normal brain development."
78. Hazlett et al., "Early brain development."
79. Reviewed in Teffer and Semendeferi, "Human prefrontal cortex." See also Buxhoeveden et al., "Reduced minicolumns"; Courchesne et al., "Neuron number and size"; Morgan et al., "Abnormal microglial-neuronal spatial organization"; Morgan et al., "Microglial activation." Unfortunately, most studies of cortical microstructure in autism do not include individuals with AS. For a report about cortical microstructure in two Aspies, see Casanova et al., "Asperger's syndrome and cortical neuropathology," *J Child Neurol* 17 (2): 142–45.
80. Courchesne et al., "Normal brain development."

81. Courchesne et al., "Neuron number and size."
82. Sakai et al., "Fetal brain development."
83. A prenatal origin for ASD is also suggested by a 2014 study by Daniel Campbell and colleagues of 200 toddlers with ASD compared to 147 TD infants. At birth, infants who would later develop ASD and TD infants showed no significant difference in brain size. However, the average gestational age of the ASD newborns was nearly a full week shorter than that for TD infants, which was highly statistically significant (Campbell, Chang, and Chawarska, "Early generalized overgrowth," 1065.) This makes sense in terms of the obstetrical dilemma, that is, babies' heads cannot exceed a certain size if they (and perhaps their mothers) are to survive birth. ASD fetuses get to that size before TD fetuses, which suggests the ASD fetuses are riding the top of the brain-growth curve during prenatal development.
84. The solid part of the ASD curve in figure 1.5 is based on a study that included one hundred males with ASD whose full-scale IQs ranged from 49 to 137 (Lange et al., "Longitudinal volumetric brain changes"). Although the authors did not recognize AS as a distinct form of autism, the IQ range in the group of autists they studied strongly suggests that it included (so-called) low-functioning, high-functioning, and AS individuals (Chiang et al., "Meta-analysis").
85. Developmental, Disabilities Monitoring Network Surveillance Year, and Principal Investigators, "Prevalence of autism spectrum," 17 (Table 4). This study was based on a 2010 survey of 363,749 eight-year-old children from eleven American sites in different states. One in 68 (about 1.5 percent) were diagnosed with ASD. Among the 3,822 of the children who were diagnosed with a specific subtype of autism, the percentage diagnosed with AS ranged from 7 to 17 percent, with an average of 11 percent. A 2016 update for 2012 also reported a prevalence for ASD of 1 in 68 among eight-year-old children (Christensen et al., "Prevalence and characteristics").
86. Gillberg and de Souza, "Head circumference in autism," 299.
87. Dementieva, "Accelerated head growth," 102. This study appears to have focused on individuals with Autistic Disorder (based on *DSM-IV*), but the extent to which individuals with AS were or were not included is not clear.
88. Pugliese et al., "Anatomy of extended limbic pathways," 433.
89. DeSilva, "Brains, birth, bipedalism"; Passingham, "Changes in the size"; Schultz, "Relative size of the cranial capacity."
90. Schumann et al., "Longitudinal magnetic resonance imaging."
91. Psychiatrist John Rubenstein of the University of California, San Francisco, was one of the first researchers to recognize that evolutionarily driven expansion of the brain may have contributed to autism (Rubenstein, "Three hypotheses for developmental defects").
92. Needless to say, this idea runs counter to conventional concepts of normality and abnormality, which reflect the degree to which individual traits and behaviors coincide with, or deviate from, average.

93.   Jaarsma and Welin, "Autism as a natural human variation," 25.

Chapter 2

1.   The opening epigraph is from Asperger, "Autistic psychopathy in childhood," 80.
2.   American Psychiatric Association, *DSM-IV*.
3.   Nevertheless, one study found that sensory abnormalities were a pronounced problem in over 90 percent of children and adults with all kinds of autism, "a striking finding given that sensory problems are not an essential diagnostic criterion for autism" (Leekam et al., "Describing the sensory abnormalities," 906). Although the 2013 revision of the Diagnostic and Statistical Manual (*DSM-5*) lists hyper- or hyporeactivity to sensory stimuli or unusual interest in sensory experiences as a possible symptom of Autism Spectrum Disorder, sensory criteria are still not part of the core requirements.
4.   Coren, Porac, and Ward, *Sensation and perception*, 2.
5.   Myles et al., "Children with Asperger syndrome."
6.   Ibid., 288, 287.
7.   Markram, Rinaldi, and Markram, "Intense world syndrome," 92.
8.   Myles et al., "Children with Asperger syndrome."
9.   Falk, "Adaptive value of happiness."
10.   Simone, *Aspergirls*, 24.
11.   Chamak et al., "What can we learn." Nine of the twenty-one individuals had been diagnosed with AS. Because language skills are needed to write books, it's a good bet that the rest of the authors had HFA or undiagnosed AS.
12.   Ibid., 271.
13.   Quoted by Chamak et al., "What can we learn," 274.
14.   Simone, *Aspergirls*, 45.
15.   Ibid., 48. No matter how soothing or pleasurable it might be, however, stimming can sometimes be taken to extremes. For example, Aspie writer Liane Holliday Willey writes: "When I am stressed too much, my body seems to evaporate. It's like my body slips into a deep meditative state while the rest of me flakes off to look around the room and take in every detail . . . Then something in me clicks. When my body gets too separated from my conscious movements and thoughts, I instinctively go to desperate lengths to ground myself. When my father died, I had this separation effect. . . . The only thing I could think to do to reconnect was to scrape. I took a key out of my pocket and started raking it across my arm. . . . To someone like me who has hyposensitive sensory integration dysfunction, self-injury feeds the need to feel. . . . Truly, without a wake-up call to my sensory system, my body would have shut down. If only this were a safe way to patch things up. I understand why girls cut" (Willey, *Skills for Asperger women*, 94–95).
16.   Grandin, *Thinking in pictures*, 63.

17. Grandin and Scariano, *True story*, 30–31.

18. Jackson, *Freaks, geeks and Asperger syndrome*, 107–8.

19. Ibid., 66.

20. Hall, *Asperger syndrome*, 48–50.

21. Tammet, *Born on a blue day*, 62.

22. Simone, *Aspergirls*, 39.

23. Grandin, *Thinking in pictures*, 63.

24. Blakemore et al., "Why can't you tickle yourself?"

25. Ashwin et al., "Eagle-eyed visual acuity."

26. Bonnel et al., "Enhanced pitch sensitivity."

27. Green et al., "Overreactive brain responses to sensory stimuli." Given Aspies' hypersensitivity to loud noises and bright lights, it is not surprising to learn that primary auditory and visual cortices of HFA/AS youths respond more to such stimuli than is the case for TD individuals, as do emotion processing regions in their brains (amygdala, hippocampus, and prefrontal cortex).

28. Robinson, *Be different*, 181.

29. Simone, *Aspergirls*, 37.

30. Attwood, *Guide to Asperger's syndrome*, 288.

31. Willey, *Skills for Asperger women*, 93. It is worth noting, however, that on page 94 Willey also remarks that she is not completely "immune to pain" and that she takes medication for migraines.

32. Squire et al., *Fundamental neuroscience*, 602.

33. Sainsbury, *Martian in the playground*, 23.

34. Squire et al., *Fundamental neuroscience*. The firing rate of a receptor signals the strength of its particular sensory input, so an individual's perception about the intensity of a specific sensation results from the number of neurons involved and their firing rates. A separate group of receptors may inform the brain only about light pressure, whereas another group sends information specific to tissue damage. Alteration in the pathways for either one can alter the perception they regulate (one way or another).

35. Riquelme, Hatem, and Montoya, "Abnormal Pressure Pain." According to another study (Kaiser et al., "Brain mechanisms"), autistic individuals are hyposensitive to social touch (such as a gentle brush of the arm) in parts of the brain that are involved with social-emotional processing but hypersensitive to touch elsewhere in the brain.

36. Ebisch et al., "Altered intrinsic functional connectivity."

37. Hardan et al., "Brief report"; McAlonan et al., "Distinct patterns of grey matter abnormality."

38. Brown, "Vestibular sense."

39. The three inner-ear canals are called the anterior, lateral, and posterior semicircular canals, which are stimulated, respectively, by rotation that is diagonally forward, horizontal, and diagonally backward. Together, the

semicircular canals in both ears provide ongoing information about the velocity and angles of the head's movements.

40. Partygoers sometimes experience temporary jittery eyes (nystagmus) from consuming too much alcohol in too little time. When they lie down, a sense of vigorous spinning can occur because of the alcohol's effects on the receptors in the semicircular canals (Squire et al., *Fundamental neuroscience*, 702). Sometimes affected persons are able to suppress the response by "grounding" themselves by placing a foot on the floor and/or by fixating on a stationary point in the room.

41. Ornitz, "Modulation of sensory input"; White, "Abnormal vestibulo-ocular reflexes."

42. Ornitz, "Modulation of sensory input." According to Ornitz (197), because the vestibular system modulates both "motor output at the time of sensory input and sensory input at the time of motor output," the extreme suppression of eye movements in response to both vestibular (rotation, bed rocking) and visual (light) sensory stimulation in autistic children represents a general dysfunction in the integration of sensory and motor systems.

43. Tammet, *Born on a blue day*, 25.

44. Grandin, *Thinking in pictures*, 45.

45. Grandin and Scariano, *True story*, 18.

46. Ibid., 72. Although many Aspies like to spin, it is important to clarify that not all of them have such user-friendly vestibular systems. For example, Liane Holliday Willey noted, "Motion is not my friend. My stomach tips and spills when I look at a merry-go-round, or drive my car over a hill or around a corner too quickly" (Willey, *Pretending to be normal*, 76).

47. Tammet, *Born on a blue day*, 22.

48. Grandin and Scariano, *True story*, 18–19.

49. Brown, "Forgotten sense." Pressure receptors in the soles of the feet add information about the position of the body with respect to gravity, and nerves in the skin contribute to the sense of position (Squire et al., *Fundamental neuroscience*, 589).

50. Brown, "Forgotten sense," 23. Brown suggests a variety of therapeutic measures that may lessen these symptoms, including deep pressure massage, weighted clothing or bedcovers, chewing gum, and others.

51. Ornitz, "Modulation of sensory input," 199.

52. Ibid., 200.

53. Ozonoff et al., "Atypical object exploration."

54. Receptors in the skin help relay information about form and texture, grip, vibration, hand shape, and position of the limbs. According to Squire et al., *Fundamental neuroscience*, 583, four specific kinds of mechanosensory receptors in the skin contribute to proprioception, which would have been a radical revelation a decade ago. See their section titled "Proprioception involves more than proprioceptors" (589).

55. Weimer et al., "'Motor' impairment in Asperger syndrome," 99. The authors found Aspie males to be significantly poorer than similarly aged neurotypical males on tests of proprioception, including standing on one leg with the eyes closed, walking heel-to-toe in a straight line, touching the finger to the thumb as many times as possible in a set amount of time, and assuming certain postures with the entire body (such as marching in place like a soldier).

56. Eibl-Eibesfeldt, *Human ethology*, 209.

Chapter 3

1. The opening epigraph is from Asperger, "Autistic psychopathy in childhood," 79.

2. Ghaziuddin, "Defining the behavioral phenotype." Consistent with *DSM-IV*'s criteria for AS, the author identified autistic people with IQs over 70 who had strong isolated interests but lacked language delay as having AS, and autists with a similar lower limit to their IQs who experienced language delay as HF.

3. Ghaziuddin, "Defining the behavioral phenotype," 141. Ghaziuddin subsequently provided a fuller list of features that distinguish AS from HFA. In addition to "quality of social impairment (active but odd rather than aloof and passive)," it includes "idiosyncratic interests (often sophisticated and intellectual); and communication style (often pedantic and verbose)" (Ghaziuddin, "Drop Asperger syndrome?" 1147).

4. Rueda, Fernández-Berrocal, and Schonert-Reichl, "Empathic abilities and theory of mind."

5. Premack and Woodruff, "Does the chimpanzee."

6. Slaughter, "Theory of mind in infants."

7. Baron-Cohen, Leslie, and Frith, "Does the autistic child have a 'theory of mind'?"

8. Paynter and Peterson, "Language and ToM."

9. Ibid., 383.

10. Spek, Scholte, and Van Berckelaer-Onnes, "Theory of mind in adults."

11. Paynter and Peterson, "Language and ToM."

12. Ibid., 383.

13. Kaland et al., "New 'advanced' test."

14. Ibid., 519.

15. Ibid.

16. Ibid.; Kuroda et al., "New advanced 'mind-reading' tasks."

17. Baron-Cohen and Wheelwright, "Empathy quotient."

18. Bloom, "Against empathy"; Dvash and Shamay-Tsoory, "Theory of mind"; Walter, "Social cognitive neuroscience."

19. Dvash and Shamay-Tsoory, "Theory of mind"; Walter, "Social cognitive neuroscience."

20. Iacoboni, "Imitation, empathy, and mirror neurons"; Dapretto et al., "Understanding emotions in others."

21. Baron-Cohen and Wheelwright, "Empathy quotient." Unfortunately, few if any studies compare empathy in individuals with AS and HFA (for review, see Rueda, Fernández-Berrocal, and Schonert-Reichl, "Empathic abilities and theory of mind").

22. Montgomery et al., "Emotional intelligence in Asperger"; Montgomery et al., "Young adults with Asperger."

23. Montgomery et al., "Emotional intelligence in Asperger," 577.

24. Klin, "Attributing social meaning," 841.

25. Meltzoff and Decety, "What imitation tells us," 491.

26. Meltzoff, "Origins of theory of mind."

27. Meltzoff, "Like me," 130.

28. Byrne and Whiten, *Machiavellian intelligence*.

29. De Waal, "Putting the altruism back into altruism." The author believes empathy evolved in hominins because it facilitated adaptive behaviors known as reciprocal altruism ("You scratch my back and I'll scratch yours") and kin selection (helping one's relatives increases the probability that a greater number of copies of one's own genes will be selected for).

30. Paynter and Peterson, "Language and ToM." Astington and Jenkins, "Language and theory-of-mind development."

31. For example, this idea is supported by the fact that the neurological substrates for the relatively late emergence of symbolic arithmetic in children seem to be influenced by the neurological architecture underlying their earlier development of language (Pinel and Dehaene, "Beyond hemispheric dominance").

32. Seeley et al., "Dissociable intrinsic connectivity networks."

33. Sestieri et al., "Domain-general signals in the cingulo-opercular network."

34. Hofman, "When bigger is better"; Bullmore and Sporns, "Economy of brain network organization."

35. Gazzaniga, "Forty-five years of split-brain research."

36. Marked brain lateralization in humans is consistent with the evolutionary trend for an increased number of functional areas that process information locally, which prevents an excessive number of long-distance fibers from swamping the brain, thus leading to inefficient processing (Hofman, "When bigger is better").

37. Gotts et al., "Two distinct forms of functional lateralization"; Pinel and Dehaene, "Beyond hemispheric dominance"; Ouimet et al., "Auditory-musical processing in autism"; Rueckl et al., "Universal brain signature"; Watanabe et al., "Asymmetry of the endogenous opioid system."

38. Nielsen et al., "Evaluation of the left-brain vs. right-brain hypothesis"; Ouimet et al., "Auditory-musical processing in autism"; Schore, "Right brain is dominant."

39. Goldberger, "Music of the left hemisphere"; Lindell and Hudry, "Atypicalities

in cortical structure"; Saban-Bezalel and Mashal, "Hemispheric processing of idioms."

40. Although neuroscientists are currently debating the extent to which the left and right frontal lobes may process positive and negative emotions, respectively, most seem to agree than the right hemisphere is generally more involved in processing emotions than the left. For details, see Costanzo et al., "Hemispheric specialization in affective responses"; Demaree et al., "Brain lateralization of emotional processing"; Miller et al., "Issues in localization of brain function"; Pichon and Kell, "Emotional prosody generation"; Watanabe et al., "Asymmetry of the endogenous opioid system."

41. Gazzaniga, "Forty-five years of split-brain research"; Tseng et al., "Facial emotion recognition."

42. Gazzaniga, "Forty-five years of split-brain research"; Ouimet et al., "Auditory-musical processing in autism"; van der Knaap and van der Ham, "Corpus callosum mediate."

43. Nolte, *Human brain*; Thurm et al., "Neuroanatomical aspects." In fact, damage to large parts of the right parietal lobe causes difficulty with spatial orientation toward everything on the left side. Affected individuals may even ignore the halves of objects to the left, including their own bodies. Thus they may not shave the left side of their face or put a glove on their left hand.

44. Rather than observing neuronal firing per se, fMRI detects the changes that support neural activity in resting or active brains by measuring the proportion of oxygenated/deoxygenated hemoglobin in the momentary blood supply of different parts of the brain. A greater BOLD (blood oxygen level dependent) signal in a small region indicates that it is busy, or activated, whereas a lower signal shows it is less so (relatively "deactivated"). Among other techniques, the BOLD signal in a specific small region (called a seed) can be compared to the signals in all other brain regions to detect the ones that are synchronized with its particular activity pattern. Together, the correlated regions of a given seed form a "functionally connected" network, even though its components are distributed among different parts of the brain. Another very powerful method of analyzing fMRI data, called independent component analysis, automatically determines connected networks (resting or active) across the whole brain without relying on the use of seeds. For details, see Smith et al., "Correspondence of the brain's functional architecture."

45. Buckner, Andrews-Hanna, and Schacter, "Brain's default network"; Kennedy and Courchesne, "Intrinsic functional organization."

46. Maximo, Cadena, and Kana, "Implications of brain connectivity."

47. Smith et al., "Correspondence of the brain's functional architecture."

48. Damoiseaux et al., "Consistent resting-state networks"; Seeley et al., "Dissociable intrinsic connectivity networks"; Smith et al., "Correspondence of the brain's functional architecture."

49. Smith et al., "Correspondence of the brain's functional architecture," 13043.

Curiously, only two of these twenty networks appear to be strongly lateralized and are largely left-right mirrors of each other that correspond to language functions on the left and perception of the body and pain on the right. "Given the known lateralization of language function, it is not surprising that these (mirrored) networks have such different behavior domain associations" (Smith et al., "Correspondence of the brain's functional architecture," 13042).

50. Buckner, Andrews-Hanna, and Schacter, "Brain's default network"; Di Martino et al., "Autism brain imaging data"; Elton and Gao, "Task-positive functional connectivity"; Fox et al., "anticorrelated functional networks"; Kennedy and Courchesne, "Functional abnormalities"; Kennedy and Courchesne, "Intrinsic functional organization"; Mars et al., "Default mode network"; Molnar-Szakacs, Uddin, and Uddin, "Self-processing"; O'Callaghan et al., "Shaped by our thoughts."

51. Spreng, "Fallacy."

52. Dunbar and Shultz, "Evolution in the social brain."

53. Elton and Gao, "Task-positive functional connectivity."

54. Tamir et al., "Reading fiction and reading minds."

55. Gilbert and Wilson, "Prospection," 1354. See also Killingsworth and Gilbert, "Wandering mind."

56. Buckner, Andrews-Hanna, and Schacter, "Brain's default network"; Di Martino et al., "Autism brain imaging data"; Fox et al., "Anticorrelated functional networks"; Lynch et al., "Default mode network."

57. Buckner, Andrews-Hanna, and Schacter, "Brain's default network"; Mars et al., "Default mode network."

58. Dvash and Shamay-Tsoory, "Theory of mind"; Lynch et al., "Default mode network."

59. Lynch et al., "Default mode network."

60. Bzdok et al., "Characterization of the temporo-parietal junction."

61. Ibid. Regarding toggling between internal and external mindsets, see also Broyd at al., "Default-mode brain dysfunction." All three main hubs of the default network shown in figure 3.2 are functionally connected to each other as well as to many other subnetworks.

62. Kennedy and Courchesne, "Intrinsic functional organization."

63. Barttfeld et al., "State-dependent changes of connectivity patterns."

64. Kennedy and Courchesne, "Intrinsic functional organization," 1883.

65. Uddin et al., "Salience network–based classification," 871.

66. Ibid., 874.

67. Nomi and Uddin ("Changes in large-scale network") speculate that the hyperconnectivity in resting-state neurological activity of HFA/AS children may have resulted from their early brain spurts and then subsided over time in association with an atypical amount of pruning of neuronal connections (chapter 1). The authors think atypical within- and between-network functional connectivity of the default mode network may be a possible brain

marker of ASD (739). See also Rane et al., "Connectivity in autism"; Williams et al., "Brain function differences."

68. Di Martino et al., "Functional brain correlates"; Singer, Critchley, and Preus-choff, "Common role of insula."

69. Uddin et al., "Salience network–based classification," 869.

70. Cascio et al., "Affective neural response."

71. Barttfeld et al., "State-dependent changes of connectivity patterns."

72. Ibid., 3660.

73. Ibid.

74. Although the autistic individuals in the study were confined to Aspies, the authors referred to that group as HFA to be consistent with *DSM-5*.

75. Mueller et al., "Multimodal MRI study." White matter is fatty tissue that coats the fibers (axons) that conduct impulses away from cell bodies (neurons), whereas gray matter is tissue that contains a preponderance of neurons. In addition to resting-state functional connectivity MRI, the authors used diffusion tensor imaging, which determines the degree of orderliness ("structural integrity") in white-matter fiber tracts that connect different regions, and voxel-based morphometry to compute regional differences in volumes of gray matter.

76. Bzdok et al., "Characterization of the temporo-parietal junction."

77. This is supported by the finding in another study that activity in the right TPJ was increasingly reduced in a group of twenty-nine adult male Aspies in association with the severity of their ability to mentalize about themselves and others (mindblindness) (Lombardo et al., "Specialization of right temporo-parietal junction").

78. Murphy et al., "Anatomy and aging of the amygdala and hippocampus"; Nordahl et al., "Increased rate of amygdala growth in children."

79. Interestingly, the white matter in certain pathways of the limbic system is also significantly different in people with AS and TD individuals, including under-development of a bundle that connects the right frontal and temporal lobes in the former (Pugliese et al., "Anatomy of extended limbic pathways").

80. Welchew et al., "Functional disconnectivity."

81. Green et al., "Overreactive brain responses to sensory stimuli"; Tottenham et al., "Elevated amygdala response."

82. Whyte et al., "Animal, but not human." Although this article does not distinguish AS from HFA, the data provided for subjects strongly suggest that Aspies are well represented.

83. McAlonan et al., "Differential effects on white-matter systems."

84. Lau et al., "Autism traits in individuals"; Schipul, Keller, and Just, "Inter-regional brain communication"; Vidal et al., "Mapping corpus callosum."

85. McAlonan et al., "Differential effects on white-matter systems." Aspies also had more white matter near the left TPJ. In keeping with this, the intra-parietal sulci are also deeper in Aspies than in TD individuals (Nordahl

et al., "Cortical folding abnormalities"). In earlier work, McAlonan and
her colleagues found a similar pattern in frontal lobe gray matter patterns
(McAlonan et al., "Distinct patterns of grey matter abnormality").

86. Gunter, Ghaziuddin, and Ellis, "Asperger syndrome," 279. This article sum-
     marizes research indicating that AS may be associated with right-hemisphere
     deficits, and HFA with left-hemisphere deficits. Lindell and Hudry ("Atypi-
     calities in cortical structure," 260) state a "growing body of evidence indicates
     atypical cortical lateralization in people with ASDs, with a number of key
     left-hemisphere language regions showing reduced or reversed asymmetry,"
     but suggest that, although this finding applies to individuals with HFA, it does
     not seem to be the case for Aspies. See also Bonnel et al., "Enhanced pitch
     sensitivity" for evidence that people with HFA and AS are lateralized differ-
     ently, with the former but not the latter manifesting enhanced pure-tone pitch
     discrimination.

87. Ghaziuddin and Mountain-Kimchi, "Defining the intellectual profile of
     Asperger"; Floris et al., "Atypically rightward cerebral asymmetry"; Lai et al.,
     "Neuroanatomy of individual differences."

88. Ellis et al., "Right hemisphere cognitive deficits"; Gold and Faust, "Right
     hemisphere dysfunction"; Gunter, Ghaziuddin, and Ellis, "Asperger syn-
     drome"; Kornmeier et al., "Different view on the checkerboard?"; McAlonan
     et al., "Differential effects on white-matter systems"; McKelvey et al.,
     "Right-hemisphere dysfunction"; Volkmar et al., "Asperger's disorder." How-
     ever, see Barttfeld et al., "State-dependent changes of connectivity patterns,"
     for evidence that right-lateralized regions may be significant for interocep-
     tive processing in HFA and AS.

89. Ecker et al., "Neuroimaging in autism spectrum"; Happé and Frith, "Weak
     coherence account." See also Bloemen et al., "White matter integrity," which
     speculates that widespread abnormalities in white matter indicate that Aspies
     have fewer long-range connections between separated regions than TD indi-
     viduals. However, one recent study of functional and structural networks in
     children and adolescents with HFA found both long-range and short-range
     connections to be reduced within specific networks of autistic compared to
     TD individuals, whereas connections between visual, sensorimotor, attention,
     and default networks were increased, thus favoring global over local process-
     ing (Rudie et al., "Altered functional and structural"). Interestingly, this same
     study (90) observed relatively few alterations "in the frontal attention/cogni-
     tive control network, which might reflect relatively intact cognitive skills in
     high-functioning individuals with ASD."

90. Bucaille et al., "Cognitive profile in adults with Asperger syndrome."

91. Ibid., 7.

92. It is important to note, however, that Aspies often do not manifest certain
     strengths associated with the left hemisphere, such as fine motor control of
     the right hand. Despite the apparent different patterns of brain lateralization

in HFA and AS compared to TD individuals, the right and left hemispheres appear to be less connected throughout the entire length of the corpus callosum, especially in anterior regions, in both groups (Vidal et al., "Mapping corpus callosum"). This finding, along with the discovery that very young sleeping toddlers who were eventually diagnosed with autism show extremely weak functional connectivity between the two hemispheres (Dinstein et al., "Disrupted neural synchronization in toddlers"), suggests that the early brain spurt in autistic infants may entail atypical development of the corpus callosum and, perhaps, brain lateralization. (Regarding neurotypical development of the corpus callosum, see Gilmore et al., "Early postnatal development of corpus collosum.") AS may be associated with decreased functional connectivity, not only across the midline of the brain, but also between parts of the frontal and parietal association cortices (Just et al., "Functional and anatomical cortical underconnectivity").

93. Grandin, *Thinking in pictures.*
94. Hofman, "When bigger is better."
95. Asperger, "Autistic psychopathy in childhood," 75; Grigorenko, Klin, and Volkmar, "Hyperlexia."
96. Moseley et al., "Brain routes for reading." Remarkably, Aspie adults differentially engage dorsal parietal lobe and frontal speech circuits when reading, whereas TD adults recruit more ventral occipito-temporal and rostral frontal routes.
97. Baron-Cohen, "Autism and the technical mind."
98. Liu, Shih, and Ma, "Are children with Asperger syndrome creative."
99. Happe and Vital, "What aspects of autism predispose to talent?"
100. Barttfeld et al., "State-dependent changes of connectivity patterns."
101. Csikszentmihalyi, *Flow.*
102. Dhiman, "Mindfulness and the art of living creatively," 26.
103. Smith, "Neuroscience of spiritual experience."
104. Sauer, Walach, and Schmidt, "Assessment of mindfulness."
105. Farb et al., "Attending to the present."
106. Ibid.
107. Spek, van Ham, and Nyklíček, "Mindfulness-based therapy."
108. Ulrich et al., "Neural correlates," 199.
109. Dahlin, Taylor, and Fichman, "Today's Edisons"; Schoenmakers and Duysters, "Technological origins of radical inventions."
110. Singh and Fleming, "Lone inventors," 44. Interestingly, the "inventions" of nonhuman primates also seem to be the result of one or a few individuals, such as the invention of protocultural practices of Japanese macaques, like washing wheat and salting sweet potatoes.
111. Hurlburt, Happe, and Frith, "Sampling the form of inner experience."
112. As noted in the introduction, Eve's British spellings are retained in her parts of the book.

Chapter 4

1.   The opening epigraph is from Asperger, "Autistic psychopathy in childhood," 84.
2.   Attwood, "Girls with Asperger's syndrome," 1.
3.   Baron-Cohen et al., "Attenuation of typical sex differences"; Geier et al., "Evaluation of the role and treatment."
4.   Frazier et al., "Behavioral and cognitive characteristics"; Werling and Geschwind, "Sex differences in autism."
5.   Goin-Kochel et al., "Lack of evidence."
6.   Attwood, *Complete guide to Asperger's*; Attwood, "Girls with Asperger's syndrome"; Kopp and Gillberg, "Girls with social deficits"; Lai et al., "Male and female adults"; Rivet, Taylor, and Matson, "Gender differences in core"; Szatmari et al., "Sex differences in repetitive stereotyped behaviors."
7.   Lai et al., "Male and female adults."
8.   Attwood, *Complete guide to Asperger's*; Attwood, "Girls with Asperger's syndrome"; Ernsperger and Wendel, *Girls under the umbrella*; Gillberg, *Guide to Asperger syndrome*; Simone, Aspergirls; Willey, *Pretending to be normal*.
9.   Lai et al., "Male and female adults," 6.
10. Rynkiewicz et al., "Investigation of the 'female camouflage effect.'" Interestingly, using a unique computer-based method, this study found that girls with ASD used longer gestures in shorter times than male counterparts, which may be perceived by diagnosticians as more energetic and vivid non-autistic traits.
11. Rynkiewicz et al., "Investigation of the 'female camouflage effect,'" 665.
12. Beacher et al., "Autism attenuates sex differences"; Lai et al., "Male and female adults"; Iossifov, et al., "Low load for disruptive mutations."
13. As the reader is probably aware, the term "sex" refers to biological aspects of being male or female, whereas "gender" encompasses the cultural expectations and roles associated with masculinity and femininity (i.e., the social construction of sex). The latter is subject to more variation across the world than the former.
14. Ellis, "Identifying and explaining"; Hyde, "Gender similarities hypothesis."
15. Baron-Cohen, *Essential difference*; Baron-Cohen, "Testing the extreme male brain." In other words, Baron-Cohen has shown that more TD males than females show a profile of systemizing skills being stronger than empathizing skills, and vice versa.
16. Peters and Battista, "Mental rotation stimulus library"; Peters et al., "Redrawn Vandenberg and Kuse mental"; Shepard and Metzler, "Rotation of three-dimensional objects."
17. Stoet, Gijsbert, and Geary, "Sex differences in academic."
18. Lippa, "Sex differences in personality traits," 619.
19. Ibid., 634.

20. Schmitt, "Evolution of culturally-variable sex differences"; Vandermassen, "Feminist Darwinian perspective"; Willey et al., "Mating life of geeks."

21. Vandermassen, "Feminist Darwinian perspective," 733. Vandermassen's observation also holds for ideologies that label scientific findings regarding sex differences in brains as "neurosexist" (Fine, "His brain, her brain?") or as contributing to the "heterosexualization of the new autistic subject" (Willey et al., "Mating life of geeks," 370).

22. Ellis, "Identifying and explaining"; Smuts, "Male aggression against women"; Vandermassen, "Feminist Darwinian perspective."

23. Ellis, "Identifying and explaining"; Schmitt, "Evolution of culturally-variable sex differences." An interesting and counterintuitive twist documented by Schmitt is that sex differences in many traits involving personality, sexuality, and cognition are usually larger in cultures with more egalitarian sex-role socialization and gender equity. This is not always the case, however, as shown by the positive impact of access to education on women's relative to men's math scores in at least one cross cultural study (Else-Quest, Hyde, and Linn, "Cross-national patterns of gender differences in mathematics") and findings from another study (Lippa, "Sex differences in personality traits").

24. Trivers, "Parental investment and sexual selection."

25. Lancaster, "Evolutionary perspectives on sex."

26. Evans, Neave, and Wakelin, "Vocal characteristics and body."

27. Ibid.

28. Apicella, Feinberg, and Marlowe, "Voice pitch predicts reproductive."

29. Joseph, "Evolution of sex differences," 57.

30. Lancaster, "Evolutionary perspectives on sex."

31. Cahill, "Why sex matters." The tendency of some researchers to dismiss small but statistically significant sex differences as "negligible" is special pleading. "Statistically significant" means that, whatever the size of the measured difference, it is significant, not negligible.

32. Falk et al., "Sex differences in brain/body." Interestingly, brain size in rhesus monkeys also scales the same way, which I suspect has something to do with males requiring extra brain volume for visuospatial processing.

33. Luders et al., "Why sex matters"; Sowell et al., "Sex differences in cortical thickness."

34. Cosgrove, Mazure, and Staley, "Evolving knowledge of sex differences in brain."

35. Luders et al., "Why sex matters." In an elegant piece of research, these authors have settled the tricky controversy about whether differences in the volumes of certain brain components in men and women are due primarily to the different average (but overlapping) sizes of their whole brains (i.e., to geometrical ["allometric"] scaling factors) as some have suggested (Leonard et al., "Size matters: cerebral volume") rather than to biological sex differences. Contrary to much of the literature, Luders and her colleagues found that equivalently

sized brains of men and women did not differ in the total amounts of their gray and white matter. However (and importantly) women but not men did have larger amounts of gray matter in specific parts of the brain including the left orbito-frontal cortex, the left superior temporal gyrus, and the left superior frontal gyrus. These results are especially interesting in light of the role of the left hemisphere in language, and the fact that this is something at which females excel on average compared to males. Although scaling factors contribute somewhat to observed differences between men and women's brains, Luders et al. provide convincing evidence (14269) "that the observed increased regional GM [gray matter] volumes in female brains constitute sex-dependent redistributions of tissue volume (rather than individual adjustments to brain size)."

36. Allen et al., "Sexual dimorphism and asymmetries"; Sowell et al., "Sex differences in cortical thickness."

37. Among other sex differences, women have larger average hippocampi and caudate nuclei, whereas men average larger amygdalae and hypothalami. See Cosgrove, Mazure, and Staley, "Evolving knowledge of sex differences in brain"; Ingalhalikar et al., "Sex differences"; Ruigrok et al., "Meta-analysis of sex differences"; van den Bos, "Sex matters"; According to van den Bos, sex differences in the amygdala and insula may be related to well-known sex differences in coping with stress.

38. Joel et al., "Sex beyond the genitalia."

39. Indeed, it would be surprising if there were not enormous variation and overlap between the brains and behaviors of the sexes for the simple reason that genetic material from 22 of the 23 pairs of human chromosomes (the sex chromosomes being the exception) is transmitted from both parents to male and female offspring.

40. Tomasi and Volkow, "Functional connectivity."

41. Ingalhalikar et al., "Sex differences," 824. The study examined whole brains of 949 males and females ranging in age from eight to twenty-two years. See also Gong et al., "Age- and gender-related differences."

42. See also Gong et al. who note that, after controlling for brain size, "women showed greater overall cortical connectivity and the underlying organization of their cortical networks was more efficient, both locally and globally" ("Age- and gender-related differences," 15684).

43. Ingalhalikar et al., "Sex differences," 826.

44. Butler et al., "Sex differences in mental rotation."

45. Gong et al., "Age- and gender-related differences"; Kanaan et al., "Gender differences"; Tomasi and Volkow, "Functional connectivity."

46. Tomasi and Volkow, "Functional connectivity." Men also have thinner cortices in right parietal and temporal regions, which likely enhances computational speed and efficiency in visuospatial processing (Sowell et al., "Sex differences in cortical thickness").

47. Joel et al., "Sex beyond the genitalia."
48. Baron-Cohen, "Extreme male brain theory"; Ellis, "Identifying and explaining"; Schmitt, "Evolution of culturally-variable sex differences."
49. Auyeung, et al., "Fetal testosterone predicts."
50. Ibid., 4.
51. Baron-Cohen, "Autism and the technical mind," 74.
52. Baron-Cohen and Wheelwright, "Empathy quotient"; Baron-Cohen et al., "Attenuation of typical sex differences."
53. Baron-Cohen, Knickmeyer, and Belmonte, "Sex differences in the brain."
54. Baron-Cohen, "Testing the extreme male brain," 81.
55. Baron-Cohen et al., "Attenuation of typical sex differences."
56. Hines, "Prenatal endocrine influences."
57. Bejerot and Eriksson, "Sexuality and gender role in autism"; Bejerot et al., "Extreme male brain revisited." The authors included discussion of evolution in the second article.
58. Bejerot et al., "Extreme male brain revisited." 122. Bejerot and her colleagues also found gender-atypical patterns when it came to gender roles and sexuality in both autistic men and women (Bejerot and Eriksson, "Sexuality and gender role in autism").
59. Ingudomnukul et al., "Elevated rates of testosterone-related disorders."
60. Sobanski et al., "Further evidence for a low body weight." These findings for Aspies are consistent with the fact that body mass indices of male, but not female, children with autism are significantly lower than those of age-matched TD individuals (Mouridsen, Rich, and Isager, "Body mass index").
61. Schwarz et al., "Sex-specific serum biomarker patterns."
62. Ibid., 1215.
63. Although Schwarz et al. studied only twenty-two male and twenty-three female Aspie adults, their results are generally consistent with a recent study of AQ, EQ, and SQ in a huge number of TD and autistic persons across the spectrum, which found that "the cognitive profiles of both males and females with autism are shifted towards and beyond the typical male-distribution, and normative sex differences in these profiles are attenuated in autism" (Baron-Cohen et al., "Attenuation of typical sex differences," 8). The latter study, however, also found that sex differences among people on the spectrum were in the same direction as those of TD individuals, contrary to Schwarz et al.'s research that compared just Aspies to the general population. This underscores the importance of comparing specific rather than mixed populations of autistic individuals to each other and to TD individuals, in addition to the need to pay attention to sex differences/similarities.
64. Beacher et al., "Autism attenuates sex differences."
65. Ibid., 88.
66. Ibid. According to the supplemental on-line table that accompanies this article, Aspie women have somewhat better structural integrity and conduction

efficiency of the fibers passing through the body of the corpus callosum than their male counterparts, although the difference does not reach statistical significance. Nevertheless, this difference parallels the sexual dimorphism for neurotypicals described by Kanaan et al. ("Gender differences"). Together, these studies suggest Aspies and TD individuals may be similarly sexually dimorphic when it comes to the major tract that connects the two sides of the brain, consistent with Ingalhalikar et al. ("Sex differences") and figure 4.1 in the present chapter.

67. Brosnan, Daggar, and Collomosse, "Relationship between systemising and mental rotation"; Zapf et al., "Sex differences in mental rotation."

68. Butler et al., "Sex differences in mental rotation"; Hugdahl, Thomsen, and Ersland, "Sex differences in visuo-spatial processing."

69. Hugdahl, Thomsen, and Ersland, "Sex differences in visuo-spatial processing."

70. Ibid. See Peters and Battista, "Mental rotation stimulus library," for further discussion of strategies used by males and females during mental rotation.

71. Conson et al., "Motor imagery in Asperger"; Katagiri et al., "Individuals with Asperger's"; Muth, Honekopp, and Falter, "Visuo-spatial performance in autism." Silk et al., "Visuospatial processing"; Zapf et al., "Sex differences in mental rotation."

72. Katagiri et al., "Individuals with Asperger's."

73. Muth, Honekopp, and Falter, "Visuo-spatial performance in autism."

74. Falter, Plaisted, and Davis, "Visuo-spatial processing in autism"; Muth, Honekopp, and Falter, "Visuo-spatial performance in autism"; Zapf et al., "Sex differences in mental rotation."

75. As authors of visuospatial studies point out, however, their research says nothing about the validity of extreme male brain theory for other kinds of processing.

76. Beacher et al., "Sex differences and autism." Unfortunately, the more difficult tests that reveal clear sex differences are difficult to administer during MRIs (Peters and Battista, "Mental rotation stimulus library").

77. This finding was so stark, in fact, that the authors of the study concluded that, with respect to AS, "for some cognitive processes (i.e., in language domains) males and females . . . behave as a homogeneous group, whereas for others (i.e., visuospatial processing), the differential patterns of brain function hint at the validity of considering males and females as distinct sub-groups on the autism spectrum" (Beacher et al., "Sex differences and autism," 10).

78. Stoet, Gijsbert, and Geary, "Sex differences in academic."

79. Lippa, "Sex differences in personality traits."

80. Attwood, "Girls with Asperger's syndrome."

81. Attwood, *Complete guide to Asperger's*, 181.

82. Ibid.

83. Attwood, "Girls with Asperger's syndrome."

84. Ibid.
85. Attwood, *Complete guide to Asperger's*, 186.
86. Norris, *Gifted adults with Asperger*.
87. Ibid., 213.
88. Ibid., 148.
89. Dane and Balci, "Handedness, eyedness"; Escalante-Mead, Minshew, and Swee-ney, "Abnormal brain lateralization"; Floris et al., "Atypically rightward cerebral asymmetry"; Lindell and Hudry, "Atypicalities in cortical structure"; Rysstad, Langseth, and Pedersen, "Non-right-handedness." Although Escalante-Mead, Minshew, and Sweeney suggest that left-handedness occurs more often in HFA than AS individuals, the relative amounts of left-handedness in Aspie males and females appears to be an intriguing but unexplored question.
90. Rysstad, Langseth, and Pedersen, "Non-right-handedness."
91. Sperry, "Hemisphere deconnection and unity," 728.
92. Partly because she stuttered severely as a child, I've long wondered if Eve might have speech and other language functions localized on the right rather than left side of her brain. Reversed dominance for speech is a rare condition, occurring in less than 1 percent of the population, but when it does happen it is usually in individuals who are strongly left-handed (Zago et al., "Association between hemispheric specialization for language"). It would be fascinating to compare handedness in Aspie and TD males and females, and to explore language dominance in their brains.
93. Lau et al., "Autism traits in individuals"; Schipul, Keller, and Just, "Inter-regional brain communication"; Vidal et al., "Mapping corpus callosum."

Chapter 5

1. The opening epigraph is from Asperger, "Autistic psychopathy in childhood," 84.
2. Darwin, *On the origin of species*.
3. Darwin, *Descent of man*.
4. Mendel, "Versuche über pflanzenhybriden." Mendel first presented his find-ings about pea genetics at scientific meetings in 1865, six years after Darwin published *On the Origin of Species* and six years before he published *The Descent of Man*.
5. More explicitly, Mendel's initial varieties of pea plants that were pure for height were either TT (tall) or tt (short). These plants produced only one kind of unit (or gene) for height, either T or t. Thus all the crosses between purely tall and purely short plants received one unit from each parent and were thus Tt. As Mendel observed, all the Tt plants looked tall, which we now know was due to the fact that T was a dominant unit and therefore expressed when com-bined with the recessive unit, t. When Mendel subsequently crossed Tt plants, however, a fourth of the resulting plants were tt and thus looked short because they had only recessive units.

6. Watson and Crick, "Structure of deoxyribonucleic acid."

7. Classic examples are the sickle cell allele for hemoglobin, which can protect against malaria in certain environments, and the alleles for coloration in peppered moths in England that facilitate camouflage against bird predators.

8. Wood et al., "Defining the role of common variation."

9. Lesk, *Introduction to bioinformatics*, 5.

10. Smith, "Role of genetic drift."

11. The introduction of agriculture was associated with standing bodies of water, which attracted mosquitoes that infected humans with malaria.

12. Kinnison and Hendry, "Rates of contemporary microevolution," 145.

13. Developmental, Disabilities Monitoring Network Surveillance Year, and Principal Investigators, "Prevalence of autism spectrum."

14. Blumberg et al., "Changes in prevalence," 1. Similarly, across the pond, a 2003–2004 study of autism-spectrum condition in British school children ages five to nine resulted in a prevalence of just under 1.6 percent (Baron-Cohen et al., "UK school-based population study"). See Moisse ("U.S. stats show autism"), however, who suggests that autism rates may now have plateaued in the United States.

15. Grinker, *Unstrange minds*; Hansen, Schendel, and Parner, "Explaining the increase."

16. Golden, *Childhood autism and assortative mating*.

17. Keyes et al., "Cohort effects explain the increase"; Beaudet, "Preventable forms of autism?"

18. Dave and Fernandez, "Rising autism prevalence," 467.

19. Alvarez-Mora et al., "Comprehensive molecular testing"; Tick et al., "Heritability of autism spectrum."

20. Lichtenstein et al., "Genetics of autism spectrum."

21. For example, a recent meta-analysis of a large number of twin studies found that heritability estimates ranged from 64 to 91 percent (Tick et al., "Heritability of autism spectrum"). See also Geschwind, "Genetics of autism spectrum"; Lai, Lombardo, and Baron-Cohen, "Autism"; Lichtenstein et al., "Genetics of autism spectrum"; Persico and Napolioni, "Autism genetics"; Ronald and Hoekstra, "Decade of new twin studies"; Sandin et al., "Familial risk of autism."

22. Klin et al., "Approaches to Asperger syndrome."

23. Alvarez-Mora et al., "Comprehensive molecular testing"; Lai, Lombardo, and Baron-Cohen, "Autism."

24. Ronald and Hoekstra, "Decade of new twin studies.

25. Bettelheim, *Infantile autism*.

26. Kong et al., "Rate of *de novo* mutations." With respect to *de novo* mutations, the father's age may be a potentially more significant risk factor than the mother's age because females are born with all their eggs, whereas males continue to produce new sperm that are subject to new copying error mutations throughout life (Keller and Miller, "Resolving the paradox").

27. Sandin et al., "Advancing maternal age."

28. Beaudet, "Preventable forms of autism?" Hallmayer et al., "Genetic heritability and shared environmental factors."

29. Suren, et al., "Association between maternal use of folic acid."

30. Lai, Lombardo, and Baron-Cohen, "Autism"; Taylor et al., "Autism and measles."

31. Gunter, "Seeing the spectrum entire."

32. Agbese, Velott, and Leslie, "Autism and vaccines."

33. Frazier et al., "Behavioral and cognitive characteristics"; Hallmayer et al., "Genetic heritability and shared environmental factors."

34. Tick et al., "Heritability of autism spectrum." This article, which provides a thorough meta-analysis of previous twin studies, concludes that earlier reports of significant environmental influences are likely due to statistical artifacts. The view that autism is not particularly strongly influenced by shared environmental factors is also supported by Lichtenstein et al., "Genetics of autism spectrum," and Sandin et al., "Familial risk of autism."

35. Colvert et al., "Heritability of autism spectrum," 421.

36. Li et al., "Association of maternal obesity and diabetes."

37. Geschwind, "Genetics of autism spectrum"; Persico and Napolioni, "Autism genetics."

38. Lai, Lombardo, and Baron-Cohen, "Autism."

39. Geschwind, "Genetics of autism spectrum," 412.

40. Asperger, "Autistic psychopathy in childhood," 84.

41. Reviewed in McPartland and Volkmar, "Asperger syndrome."

42. Klin et al., "Approaches to Asperger syndrome." These parents and grandparents are categorized as having a "Broader Autism Phenotype," which is associated with (often milder) manifestations of autistic traits related to social, communication, and restricted behaviors and interests. See also McPartland and Klin, "Asperger's syndrome."

43. Salyakina et al., "Variants in several genomic regions." This genome-wide association study (GWAS) initially studied the entire genomes of 392 individuals from 124 families that had at least one Aspie member, and followed up with another validation study in 468 individuals from 110 other families with one or more Aspie member.

44. Salyakina et al., "Variants in several genomic regions." See also Colvert et al., "Heritability of autism spectrum"; Frith, "Controversies about Asperger syndrome."

45. Lai, Lombardo, and Baron-Cohen, "Autism."

46. Chakrabarti et al., "Genes related to sex."

47. El-Fishawy and State, "Genetics of autism." This article discusses deleterious variants of genes that influence the junctions between nerve cells in the brain in two families that had an Aspie member. In these cases, the rare *de novo* mutations happened to be on the X chromosome. Such variations

potentially affect males and females differently because they are sex-linked. A child's sex is determined by the 23rd pair of chromosomes, known as the sex chromosomes. Females have a pair of X chromosomes; males have one X and one Y chromosome. Mothers pass one of their two X chromosomes to all their offspring. Fathers provide either and X or a Y chromosome and, thus, determine the sex of their offspring. Typically, if a mutation is on a mother's X chromosome, it will be inherited by approximately half of her sons and daughters, but will be manifested only in the sons because they will not have another (nonmutated) X chromosome to mask its effects. Such mutations are potentially interesting because AS occurs much more often in males than females.

48. Iourov et al., "Asperger syndrome candidate gene."

49. Sebat et al., *De novo* copy number."

50. Chakrabarti et al., "Genes related to sex."

51. Di Napoli et al., "Oxytocin receptor (*OXTR*) gene."

52. Feldman et al., "Oxytocin pathway genes." In particular, the author summarizes studies that implicate the wider oxytocin genetic pathway (which includes several genes) in the development of mother-infant interactions, motherese vocalizations, parent-infant gaze, empathy, and theory of mind.

53. Di Napoli et al., "Genetic variant rs17225178"; Durdiakova et al., "*STX1A* and Asperger syndrome"; Durdiakova et al., "Single nucleotide polymorphism rs6716901"; Warrier, Baron-Cohen, and Chakrabarti, "Genetic variation in *GABRB3*." For discussion about the need for large sample sizes in order to detect genetic variations association with AS, see Warrier et al., "Pooled genome-wide association." See also Iourov et al., "Asperger syndrome candidate gene," and McPartland and Volkmar, "Asperger syndrome," for other references regarding the possible genetic substrates of AS.

54. Lai, Lombardo, and Baron-Cohen, "Autism," 904.

55. Iossifov et al., "Low load for disruptive mutations"; Ronemus et al., "*De novo* mutations."

56. Klei et al., "Genetic variants, acting additively"; Salyakina et al., "Variants in several genomic regions."

57. Chaste et al., "Genome-wide association study"; Gaugler et al., "Genetic risk for autism"; Klei et al., "Genetic variants, acting additively"; Lai, Lombardo, and Baron-Cohen, "Autism."

58. Lai, Lombardo, and Baron-Cohen, "Autism."

59. Nomi and Uddin, "Changes in large-scale network."

60. Chakrabarti et al., "Genes related to sex"; Di Napoli et al., "Genetic variant rs17225178"; Salyakina et al., "Variants in several genomic regions."

61. Bakken et al., "Primate brain development," 371; See also Iossifov et al., "Low load for disruptive mutations."

62. Chakrabarti et al., "Genes related to sex."

63. Geschwind, "Genetics of autism spectrum," 412.

64. Warrier et al., "Pooled genome-wide association," 2. See also Ronald and Hoekstra, "Decade of new twin studies."

65. Mouridsen et al., "Death in autism spectrum."

66. Larsen and Mouridsen, "30-year follow-up study."

67. Power et al., "Fecundity of patients."

68. Alvarez-Mora et al., "Comprehensive molecular testing."

69. Chaste et al., "Genome-wide association study"; Crespi, "Natural selection of psychosis"; Gaugler et al., "Genetic risk for autism"; Gauthier and Rouleau, "New genetic mechanism"; Keller and Miller, "Resolving the paradox"; Klei et al., "Genetic variants, acting additively"; Lai, Lombardo, and Baron-Cohen, "Autism"; Ploeger and Galis, "Evolutionary approaches to autism"; Power et al., "Fecundity of patients"; Salyakina et al., "Variants in several genomic regions"; Uher, "Role of genetic variation."

70. Lai, Lombardo, and Baron-Cohen, "Autism," 896.

71. Ploeger and Galis, "Evolutionary approaches to autism," 38.

72. Gernsbacher, Dawson, and Mottron, "Heritable, but not harmful."

73. Baron-Cohen, "Hyper-systemizing, assortative mating," 865.

74. Baron-Cohen and Hammer, "Parents of children."

75. Baron-Cohen et al., "Engineering and autism?"

76. Baron-Cohen et al., "Mathematical talent is linked."

77. Buchen, "When geeks meet."

78. Baron-Cohen, "Autism and the technical mind."

79. Silberman, "Geek syndrome," 9.

80. Roelfsema et al., "Regions in the Netherlands."

81. Asperger, "Problems of infantile autism," 49.

82. Klei et al., "Genetic variants, acting additively"; Silberman, "Geek syndrome"; Spek, Annelies, and Velderman, "Examining the relationship between Autism spectrum disorders and technical professions."

83. Baron-Cohen, "Autism and the technical mind."

84. Baron-Cohen, "Hyper-systemizing, assortative mating." Baron-Cohen quantifies levels of systemizing in AS compared to HFA, medium-functioning autism, and low-functioning autism. Unlike many researchers, in this article Baron-Cohen restricts HFA to autists who experienced language delay and have IQs at or above 85 (instead of 70).

85. Quoted by Silberman, "Geek syndrome," 12. Although the manager described the debuggers as having AS, their heavy reliance on visual thinking suggests that at least some of them may have had HFA.

86. Golden, *Childhood autism and assortative mating*.

87. Ibid., 54.

88. Ibid., 53.

89. Ibid., 4.

90. Ibid., 54.

91. Falk, "Charles Darwin." Sexual selection happens when one sex (or both)

preferentially mates and bears offspring with individuals who have certain traits. This explains, for example, the gorgeous tail feathers of peacocks, which are preferred by peahens.

92. Baron-Cohen et al., "Talent in autism."

93. Green et al., "Draft sequence of the Neandertal genome"; Meyer et al., "High-coverage genome"; Prufer et al., "Complete genome sequence of a Neanderthal."

94. Meyer et al., "High-coverage genome."

95. Hawks, "Archaic human genomes."

96. Nuttle et al., "*Homo sapiens*-specific gene family," 205.

97. Ibid., 208.

98. Recurrent duplications of the segment that contains *BOLA2* likely occurred during the exchange of genetic material ("crossing over") during the production of ova and sperm (gametes) (Nuttle et al., "*Homo sapiens*–specific gene family," 208). In unequal crossing over (suggested in ibid.), the duplicated segment would be shifted to another part of the chromosome (thus becoming a duplicate). It seems reasonable to speculate that the portion of the chromosome between the original and duplicated portions would be subject to mutations during such crossing overs.

99. Green et al., "Draft sequence of the Neandertal genome."

100. Oksenberg et al., "*AUTS2*, a gene implicated." As the author details, additional variants in this gene have also been associated with a myriad of other conditions.

101. Salyakina et al., "Variants in several genomic."

102. Krause et al, "Derived *FOXP2* variant."

103. Toma et al., "Analysis of two language-related genes."

104. Green et al., "Draft sequence of the Neandertal genome"; Vernot et al., "Excavating Neandertal and Denisovan DNA."

105. Silberman, "Geek syndrome."

Chapter 6

1. The opening epigraph is from Asperger, "Autistic psychopathy in childhood," 39.

2. Park, "Autism numbers are rising."

3. Elsabbagh et al., "Global prevalence of autism."

4. Baxter, "Epidemiology and global burden of autism."

5. This finding should not be viewed as contradicting Elsabbagh et al.'s finding of a rise in autism because the Baxter study adjusted for the kinds of variables that hypothetically accounted for the rise in the Elsabbagh study. In other words, both studies appear to attribute the apparent rise in autism to various artifacts rather than to an actual increase in the incidence of autism.

6. Baxter, "Epidemiology and global burden of autism," 607.

7. Zablotsky et al., "Estimated prevalence of autism."

8. Ibid., 3.

9. Kim et al., "Comparison of DSM-IV."

10. Kim et al., "Prevalence of autism spectrum."

11. Barbaro and Halder, "Early Identification of Autism Spectrum Disorder," 1.

12. Fombonne, "Epidemiology of pervasive developmental disorders"; Golden, *Childhood autism and assortative mating.*

13. Matson and Kozlowski, "Increasing prevalence of autism," 419.

14. Ibid., 423.

15. AS was defined in the various surveys using somewhat different diagnostic criteria. Sources included (among others) the American Psychiatric Association, *DSM-IV*; Gillberg and Gillberg, "Asperger syndrome"; and World Health Organization, *ICD-10 classification.*

16. Fombonne, "Epidemiology of pervasive developmental disorders," 592.

17. Hill, Zuckerman, and Fombonne, "Epidemiology of autism spectrum." However, one study of 4,422 eight-year-old Finnish students diagnosed in 2001–2002 cited studies that estimated frequencies of up to nearly 0.5 percent (that is, half of 1 percent of the population), just for AS (Mattila et al., "Epidemiological and diagnostic study").

18. Mattila et al.'s report of some estimates approaching nearly 0.5 percent for the prevalence of AS was consistent with a more recent study of over fifty-five thousand South Korean school children, which estimated a prevalence of 0.6 percent for AS (Kim et al., "Comparison of DSM-IV").

19. Hill, Zuckerman, and Fombonne, "Epidemiology of autism spectrum," 70.

20. Developmental, Disabilities Monitoring Network Surveillance Year, and Principal Investigators, "Prevalence of autism spectrum"; Christensen et al., "Prevalence and characteristics."

21. Extrapolating from the statistics in Table 4 of the previous reference to the entire sample of 363,749 eight-year-olds, I calculate that 587 of the 5,338 children with autism (about 11 percent) would have had AS, which translates to approximately 1 Aspie for every 619 children in the entire sample or 0.16 percent, which is only a bit higher than the estimated average of 0.12 percent (just over one-tenth of 1 percent of the population) calculated from eighteen publications by Hill, Zuckerman, and Fombonne, "Epidemiology of autism spectrum."

22. Kim et al., "Prevalence of autism spectrum."

23. Henrich, Heine, and Norenzayan, "Weirdest people."

24. Ibid., 83.

25. Ibid., 61.

26. Lotter, "Childhood autism in Africa."

27. Daley, Singhal, and Krishnamurthy, "Ethical considerations."

28. Yu and King, "Focus on autism."

29. Daley, "Need for cross-cultural research," 53.

30. Ibid.
31. Daley, Singhal, and Krishnamurthy, "Ethical considerations."
32. Ibid., 207.
33. Grinker et al., "'Communities' in community engagement," 207–8.
34. Paula et al., "Prevalence of pervasive developmental disorder in Brazil."
35. Montiel-Nava and Peña, "Epidemiological findings of pervasive developmental disorders."
36. It is noteworthy, however, that one study estimates that over 85 percent of the world's known cases of autism are from just 10 percent of the world's children who live in North America, Europe, and Japan (Barbaro and Halder, "Early Identification of Autism Spectrum Disorder").
37. Baron-Cohen, "Autism and the technical mind"; Golden, *Childhood autism and assortative mating.*
38. Russell Greaves, personal communication, March 21, 2016.
39. Marlowe, "Central place provisioning."
40. Colette Berbesque, personal communication, March 17, 2016.
41. Petković et al., "Late diagnosis of Asperger syndrome in Croatia."
42. Ibid., 427.
43. Daley, "Need for cross-cultural research," 542.
44. Baron-Cohen, "Autism and the technical mind," 75.
45. Tanidir and Mukaddes, "Referral pattern and special interests."
46. Ibid., 178.
47. Kim et al., "Prevalence of autism spectrum."
48. Norbury and Sparks, "Difference or disorder?" 51.
49. Grinker et al., "'Communities' in community engagement," 205.
50. Zhang, Wheeler, and Richey, "Cultural validity in assessment instruments."
51. Sun et al., "Exploring the underdiagnosis and prevalence."
52. Ametepee and Chitiyo, "What we know about autism in Africa"; Lotter, "Childhood autism in Africa"; Mankoski et al., "Case-series from Tanzania."
53. Falk, *Finding our tongues.*
54. Norbury and Sparks, "Difference or disorder?" 46.
55. Singer, "Why can't you be normal," 63–64.
56. Shore, *Beyond the wall,* 60.
57. Silberman, *Neurotribes.*
58. Singer, "Why can't you be normal," 66.
59. Chamak and Bonniau, "Autism and social movements"; Dalton, "Neurodiversity & HCI"; Davidson, "Autistic culture online"; van der Aa et al., "Computer-mediated communication."
60. Jaarsma and Welin, "Autism as a natural human variation."
61. Ibid., 25.
62. Ibid., 29.
63. Allred, "Reframing Asperger syndrome."

Chapter 7

1. The opening epigraph is from Asperger, "Autistic psychopathy in childhood," 88.
2. American Psychiatric Association, *DSM-5*.
3. Charlop and Haymes, "Speech and language acquisition"; Rutter, "Language disorder and infantile autism." Some sources, however, suggest that muteness may occur in as low as 25 percent of all autists (Anonymous, "Spectrum shift").
4. Although autism probably emerged after language was invented, some nonlinguistic behavioral symptoms of autism can be induced in monkeys, both behaviorally (Harlow, "Nature of love") and through genetic engineering (Liu et al., "Autism-like behaviours").
5. Gibbons, "Five matings for moderns, Neandertals," 1251; Vernot et al., "Excavating Neandertal and Denisovan DNA."
6. This conclusion is compatible with that of British archaeologist Penny Spikins, who speculates about the origins of autism by examining the archaeological record of tools and art (Spikins, *Stone age origins of autism*).
7. Silberman, *Neurotribes*, 470.
8. Day and Sweatt, "Epigenetic mechanisms in cognition"; Kanherkar, Bhatia-Dey, and Csoka, "Epigenetics across the human lifespan"; Reilly et al., "Evolutionary changes."
9. Vadée-Le-Brun, Rouzaud-Cornabas, and Guillaume Beslon, "Epigenetic inheritance speeds up."
10. Crowley and Heyer, *Communication in history*.
11. United Nations Educational, Scientific and Cultural Organization Institute for Statistics, "Adult and youth literacy."
12. Carrion-Castillo, Franke, and Fisher, "Molecular genetics of dyslexia"; Schumacher et al., "Genetics of dyslexia."
13. Soden et al., "Longitudinal stability in reading comprehension," 1.
14. Pinel et al., "Genetic and environmental influences," 13.
15. Dehaene et al., "Illiterate to literate."
16. Rueckl et al., "Universal brain signature."
17. Ibid., 15514.
18. Carrion-Castillo, Franke, and Fisher, "Molecular genetics of dyslexia"; Schumacher et al., "Genetics of dyslexia."
19. Moseley et al., "Brain routes for reading."
20. Pinel et al., "Genetic and environmental influences."
21. Moseley et al., "Brain routes for reading."
22. Dehaene and Cohen, "Recycling of cortical maps," 389.
23. Dehaene and Cohen, "Recycling of cortical maps"; Pinel et al., "Genetic and environmental influences."
24. Dehaene and Cohen, "Recycling of cortical maps."

25. Asperger, "Autistic psychopathy in childhood," 75; Grigorenko, Klin, and Volkmar, "Hyperlexia."
26. Moseley et al., "Brain routes for reading."
27. Mills, "Effects of internet use."
28. Loh and Kanai, "How has the internet."
29. Ibid., 21.
30. Wolf and Barzillai, "Importance of deep reading."
31. Loh and Kanai, "How has the internet."
32. Green and Bavelier, "Action video game modifies."
33. Loh and Kanai, "How has the internet"; Small, "Your brain on Google."
34. Maher, "Good gaming."
35. McKinlay, "Technology."
36. Ibid., 574.
37. Xu, "Towards synthetic telepathy." Don't laugh. Humans have already invented brain-computer interfaces that permit artificial limbs (prostheses) to be controlled by thought, and the first experiment of successful human brain-to-brain communication was achieved in 2013. As Xu summarizes (5), "From medicine and military technology to games and recreation, brain interfacing truly has the potential to change the world."
38. Day and Sweatt, "Epigenetic mechanisms in cognition."
39. Vadée-Le-Brun, Rouzaud-Cornabas, and Guillaume Beslon, "Epigenetic inheritance speeds up."
40. Silberman, "Geek syndrome," 7; see also Silberman, *Neurotribes*.
41. Vara, "Brain trust."
42. Golden, *Childhood autism and assortative mating*, 51.
43. Ibid., 54.
44. Baron-Cohen, "Autism and the technical mind."
45. Asperger, "Autistic psychopathy in childhood," 74.
46. Williams et al., "CAST."
47. Ozonoff et al., "Atypical object exploration."
48. Cohen et al., "Parentese prosody."
49. Ibid.; Ouss et al., "Infant's engagement and emotion"; Saint-Georges et al., "Motherese in interaction"; Tanguay, "Autism in DSM-5."
50. Myles et al., "Sensory issues in children," 287.
51. Bidinosti et al., "CLK2 inhibition"; Wright, "'CRISPR'way."
52. Tsai, "Asperger's disorder will be back," 2914.

affect. In psychology, the experience or display of a person's mood, feelings, or emotions.

allele. Alternative forms of a gene that are found at the same place on a chromosome; for example, there are three alleles (A, B, and O) for the ABO blood group gene.

Asperger syndrome. A form of autism that is associated with normal or above normal IQs and no delay in the development of language (that is, use of single words by age two and communicative phrases by age three). Compare with *high-functioning autism*.

assortative mating. Nonrandom matings based on individuals' preferences for characteristics in a mate that are similar to their own (positive assortative mating in which "likes mate with likes") or dissimilar (negative assortative mating in which "opposites attract"). Positive assortative mating is thus a form of sexual selection in which individuals with similar genetic or physical makeup mate with each other more often than expected by chance.

attachment theory. Formulated by psychoanalyst John Bowlby, this theory hypothesizes that mothers and infants evolved an adaptive need to maintain their physical proximity and emotional bonds.

axon. The extension from of a nerve cell (neuron) that conducts impulses to other cells.

balanced polymorphism. Stabalized frequencies of two alleles for the same gene in a population, despite the fact that each one is deleterious in individuals who have two copies. The alleles are maintained because of natural selection for individuals who have one copy of each (e.g., certain hemoglobin alleles that are associated with sickle cell anemia).

biocultural evolution. Evolution that results from the interaction of biological and cultural influences.

bipedalism. Walking upright on two legs, which distinguishes humans and their early relatives from other primates.

copy-number variation. Variation in the number of copies of specific sections of DNA due to deletions or duplications of relatively large portions of a chromosome. The deletion or duplication may contain numerous genes.

corpus callosum. A massive curved bundle that contains an estimated 200 million myelinated fibers that cross the midline of the brain connecting similar regions of the right and left hemispheres.

default network. The neurological network in the brain that is active when one's mind is blank or tuned to internal thoughts rather than to the external world. Its main hubs include the medial prefrontal cortex, posterior cingulate cortex, and certain lateral parietal and medial temporal regions. It communicates with subnetworks and is also called the task-negative network.

de novo **mutation.** A genetic variant that appears for the first time in an individual because of a mutation in the mother's egg, father's sperm, or fertilized egg from which the individual originated.

dyslexia. A disorder that is characterized by difficulty in learning to read.

empathy. The capacity to know and sympathize with how another individual feels and to actually experience those feelings during the process.

epigenetic. "Above genes" heritable changes in chromosomes that alter how organisms appear by turning certain genes on or off, but without changing the basic genetic composition (nucleotide sequence) of the DNA.

evo-devo. Short for "evolutionary-developmental biology," a field that compares developmental processes in different species to determine their evolutionary relationships.

"extreme male brain" theory. A theory that posits that autistic individuals have extremely masculinized brains because of prenatal exposure to high levels of testosterone.

first evo-devo trend. Delayed motor development in prehistoric hominin babies. This trend for physical late bloomers is retained in modern infants.

functional connectivity. Similar patterns of activation in separate parts of the brain, usually measured when individuals are in resting states.

functional magnetic resonance imaging. A procedure that tracks localized brain activity by measuring associated changes in the concentration of oxygen in blood.

gene pool. The set of all the genes in a population or species.

grammar. The rules for ordering components of words into words (morphology) plus the rules for ordering words and phrases into sentences (syntax) in a particular language.

grammatical speech. Speech based on acquired language-specific rules for forming words (e.g., making them past tense or plural) and organizing them into an endless variety of phrases and sentences.

gray matter. Brain tissue that is composed largely of nerve cell bodies (neurons) and short fibers that bring impulses into them.

heritability. The proportion of variation in a trait within a population that can be attributed to genes rather than environmental factors (ranging between 0, which indicates no contribution, to 1, which means all differences in the trait represent genetic variation).

high-functioning autism. An unofficial label for intellectually unimpaired autistic persons who were delayed in their development of language (did not use single words until after age two, and/or did not produce communicative phrases until after age three) and have full-scale IQs that vary among different individuals from below average to high average.

hyperlexia. Precocious, spontaneous (self-taught), and compulsive reading in young children without necessarily comprehending the meaning of words, which entails decoding words by sounding them out. Sometimes defined more generally as being able to read before starting school.

hominins. Humans and their prehistoric bipedal relatives.

limbic system. Evolutionarily old collection of brain structures deep within the brain that process emotion, motivation, learning, and memory. Includes (among other regions) cingulate and parahippocampal gyri, hippocampus, amygdala, and hypothalamus.

Machiavellian intelligence. Self-interested social or political intelligence used to navigate in or control social groups.

macroevolution. The appearance of new species, which can occur after accumulated changes over large amounts of time.

Mendelian traits. Traits controlled by dominant and recessive variations of only one gene, which follow the simple hereditary laws discovered in pea plants by Gregor Mendel. ABO blood type, sickle cell disease, and albinism are examples of Mendelian traits in humans.

microevolution. Small-scale evolution due to changes in the frequencies of genetic variants (alleles) or traits from generation to generation within a single population. Microevolution may be caused by any of a number of evolutionary forces (natural, sexual, or artificial selection; gene flow; mutation;

or random genetic drift). If it continues for long enough, microevolution may eventually result in speciation (macroevolution).

modern synthesis. Twentieth-century synthesis of Darwinian evolutionary theory with Mendel's laws of inheritance, which showed natural selection acts on genetic variation.

motherese. Baby talk or infant-directed speech. See *parentese*.

myelin. Whitish substance that insulates certain nerve fibers (axons), increasing the speed at which they conduct impulses. See *white matter*.

neurocultural evolution. Evolutionary changes in the human brain in response to cultural innovations; neurological changes, in turn, influence ongoing cultural changes. The evolution of reading is an example.

obstetrical dilemma. Difficult births due to term fetuses having heads (brains) that are too large to pass comfortably through bony birth canals that conserve morphology selected earlier in conjunction with bipedalism. Because bipedalism occurred several million years before the evolution in brain size that caused the obstetrical dilemma, its influence on the emergence of the dilemma was indirect.

parentese. Another name for *motherese*, which takes into account that fathers and others engage in baby talk.

phonemes. The smallest speech sounds that distinguish one utterance from another, e.g., the \p\ in "pad" from the \f\ in "fad."

protolanguage. The first symbolic language that emerged during hominin evolution.

putting the baby down theory. Posits that (1) because prehistoric mothers periodically put their babies down nearby, reciprocal vocal communication became elaborated between mothers and infants and led to the emergence of motherese; and (2) motherese, in turn, seeded the subsequent invention of protolanguage.

second evo-devo trend. Prehistoric infants' seeking of contact comfort from separated caregivers with evolved forms of fussing, gesturing, crying, and shedding of emotional tears; this trend is manifested in modern babies.

sexual selection. A form of natural selection in which certain traits in one sex are preferred by the other, thus increasing their prevalence in future generations. The classic example is the beautiful tail feathers of peacocks, which arose and are maintained from generation to generation because they were/are favored by peahens.

synapse. Junction where impulses pass between two nerve cells.

syntax. Rules for ordering words and phrases into proper sentences.

task-negative network. See *default network*.

task-positive network. Composed of numerous networks in the brain that are activated during tasks that require visual attention and less active when the default (task-negative) system is engaged. It includes the dorsolateral prefrontal cortex (including frontal eye fields, which help orient one's focus on particular tasks), parts of inferior parietal cortex, supplementary motor areas, and insula. The task-positive network may monitor the external environment even when the default network is active.

theory of mind. Also called mentalizing or mind reading, it refers to the capacity to infer what is in other individuals' minds, including their intentions, beliefs, desires and, in the case of affective theory of mind, feelings. See *empathy*.

third evo-devo trend. An unprecedented rapid acceleration in brain growth in prehistoric hominin fetuses and newborns; manifested in modern infants as a developmentally early "brain spurt."

white matter. Pale brain tissue that consists mostly of nerve fibers that are insulated with myelin.

Agbese, E., D. Velott, and D. Leslie. 2016. "Autism and vaccines: Are siblings affected?" Poster. International Society for Autism Research, Baltimore, Md., May 12, 2016. https://imfar.confex.com/imfar/2016/webprogram/Paper22579.html.

Allen, John S., Hanna Damasio, Thomas J. Grabowski, Joel Bruss, and Wei Zhang. 2003. "Sexual dimorphism and asymmetries in the gray-white composition of the human cerebrum." *Neuroimage* 18 (4): 880–94.

Allred, Sarah. 2009. "Reframing Asperger syndrome: Lessons from other challenges to the diagnostic and statistical manual and ICIDH approaches." *Disabil Soc* 24 (3): 343–55.

Almecija, S., J. B. Smaers, and W. L. Jungers. 2015. "The evolution of human and ape hand proportions." *Nat Commun* 6:7717.

Alvarez-Mora, Maria Isabel, Rosa Calvo Escalona, Olga Puig Navarro, Irene Madrigal, Ines Quintela, Jorge Amigo, Dei Martinez-Elurbe, Michaela Linder-Lucht, Gemma Aznar Lain, and Angel Carracedo. 2016. "Comprehensive molecular testing in patients with high functioning autism spectrum disorder." *Mutat Res-Fund Mol M* 784:46–52.

American Psychiatric Association. 1994. *Diagnostic and Statistical Manual of Mental Disorders.* 4th ed. (*DSM-IV.*) Washington, D.C.: American Psychiatric Association.

———. 2013. "ASD fact sheet." http://www.dsm5.org/Documents/Autism%20Spectrum%20Disorder%20Fact%20Sheet.pdf.

———. 2013. *Diagnostic and Statistical Manual of Mental Disorders.* 5th ed. (*DSM-5*). Washington, D.C.: American Psychiatric Association.

Ametepee, Lawrence K., and Morgan Chitiyo. 2009. "What we know about autism in Africa: A brief research synthesis." *JIASE* 10 (1): 11–13.

Anderson, M. 2014. *After phrenology: Neural reuse and the interactive brain.* Cambridge, Mass.: MIT Press.

Anonymous. 2016. "Spectrum shift." *Economist,* April 16, 2016, 18–20.

Apicella, C. L., D. R. Feinberg, and F. W. Marlowe. 2007. "Voice pitch predicts reproductive success in male hunter-gatherers." *Biol Letters* 3 (6): 682–84.

Ashwin, E., C. Ashwin, D. Rhydderch, J. Howells, and S. Baron-Cohen. 2009.

"Eagle-eyed visual acuity: An experimental investigation of enhanced perception in autism." *Biol Psychiat* 65 (1): 17–21.

Asperger, H. 1944. "Autistic psychopathy in childhood." Translated and annotated by U. Frith 1991. In *Autism and Asperger syndrome*, edited by U. Frith, 37–92. Cambridge: Cambridge University Press.

——. 1979. "Problems of infantile autism." *Commun: J Natl Autistic Soc* 13:45–52.

Astington, J. W., and J. M. Jenkins. 1999. "A longitudinal study of the relation between language and theory-of-mind development." *Dev Psychol* 35 (5): 1311–20.

Attwood, Tony. 2007. *The complete guide to Asperger's syndrome.* Philadelphia: Jessica Kingsley.

——. 2013. "The pattern of abilities and development of girls with Asperger's syndrome." http://www.tonyattwood.com.au/index. php?Itemid=181&id=80:the-pattern-of-abilities-and-development-of-girls-with-aspergers-syndrome&option=com_content&view=article.

Auyeung, Bonnie, Simon Baron-Cohen, Emma Ashwin, Rebecca Knickmeyer, Kevin Taylor, Gerald Hackett, and Melissa Hines. 2009. "Fetal testosterone predicts sexually differentiated childhood behavior in girls and in boys." *Psychol Sci* 20 (2): 144–48.

Bakken, Trygve E., Jeremy A. Miller, Song-Lin Ding, Susan M. Sunkin, Kimberly A. Smith, Lydia Ng, Aaron Szafer, Rachel A. Dalley, Joshua J. Royall, Tracy Lemon, Sheila Shapouri, Kaylynn Aiona, James Arnold, Jeffrey L. Bennett, Darren Bertagnolli, Kristopher Bickley, Andrew Boe, Krissy Brouner, Stephanie Butler, Emi Byrnes, Shiella Caldejon, Anita Carey, Shelby Cate, Mike Chapin, Jefferey Chen, Nick Dee, Tsega Desta, Tim A. Dolbeare, Nadia Dotson, Amanda Ebbert, Erich Fulfs, Garrett Gee, Terri L. Gilbert, Jeff Goldy, Lindsey Gourley, Ben Gregor, Guangyu Gu, Jon Hall, Zeb Haradon, David R. Haynor, Nika Hejazinia, Anna Hoerder-Suabedissen, Robert Howard, Jay Jochim, Marty Kinnunen, Ali Kriedberg, Chihchau L. Kuan, Christopher Lau, Chang-Kyu Lee, Felix Lee, Lon Luong, Naveed Mastan, Ryan May, Jose Melchor, Nerick Mosqueda, Erika Mott, Kiet Ngo, Julie Nyhus, Aaron Oldre, Eric Olson, Jody Parente, Patrick D. Parker, Sheana Parry, Julie Pendergraft, Lydia Potekhina, Melissa Reding, Zackery L. Riley, Tyson Roberts, Brandon Rogers, Kate Roll, David Rosen, David Sandman, Melaine Sarreal, Nadiya Shapovalova, Shu Shi, Nathan Sjoquist, Andy J. Sodt, Robbie Townsend, Lissette Velasquez, Udi Wagley, Wayne B. Wakeman, Cassandra White, Crissa Bennett, Jennifer Wu, Rob Young, Brian L. Youngstrom, Paul Wohnoutka, Richard A. Gibbs, Jeffrey Rogers, John G. Hohmann, Michael J. Hawrylycz, Robert F. Hevner, Zoltán Molnár, John W. Phillips, Chinh Dang, Allan R. Jones, David G. Amaral, Amy Bernard, and Ed S. Lein. 2016. "A comprehensive transcriptional map of primate brain development." *Nature* 535 (7612): 367–75.

Barak, M. M., D. E. Lieberman, D. Raichlen, H. Pontzer, A. G. Warrener, and J. J. Hublin. 2013. "Trabecular evidence for a human-like gait in *Australopithecus africanus*." *PLoS One* 8 (11): e77687.

Barbaro, Josephine, and Santoshi Halder. 2016. "Early identification of autism spectrum disorder: Current challenges and future global directions." *Curr Dev Disord Rep*, 1–8.

Barbeau, Elise B., Andrée-Anne S. Meilleur, Thomas A. Zeffiro, and Laurent Mottron. 2015. "Comparing motor skills in autism spectrum individuals with and without speech delay." *Autism Res* 8 (6): 682–93.

Bard, Kim A. 2004. "What is the evolutionary basis for colic?" *Behav Brain Sci* 27:459.

Baron-Cohen, Simon. 2002. "The extreme male brain theory of autism." *Trends Cogn Sci* 6 (6): 248–54.

———. 2003. *The essential difference: Men, women and the extreme male brain.* London: Penguin.

———. 2005. "Testing the extreme male brain (EMB) theory of autism: Let the data speak for themselves." *Cogn Neuropsychiatry* 10 (1): 77–81.

———. 2006. "The hyper-systemizing, assortative mating theory of autism." *Prog Neuropsychopharmacol Biol Psychiatry* 30 (5): 865–72.

———. 2012. "Autism and the technical mind: Children of scientists and engineers may inherit genes that not only confer intellectual talents but also predispose them to autism." *Sci Am* 307 (5): 72–75.

Baron-Cohen, Simon, Emma Ashwin, Chris Ashwin, Teresa Tavassoli, and Bhismadev Chakrabarti. 2009. "Talent in autism: Hyper-systemizing, hyper-attention to detail and sensory hypersensitivity." *Philos Trans R Soc Lond B Biol Sci* 364 (1522): 1377–83.

Baron-Cohen, S., S. Cassidy, B. Auyeung, C. Allison, M. Achoukhi, S. Robertson, A. Pohl, and M. C. Lai. 2014. "Attenuation of typical sex differences in 800 adults with autism vs. 3,900 controls." *PLoS One* 9 (7): e102251.

Baron-Cohen, Simon Hammer, and Jessica Hammer. 1997. "Parents of children with Asperger syndrome: What is the cognitive phenotype?" *J Cogn Neurosci* 9 (4): 548–54.

Baron-Cohen, S., R. C. Knickmeyer, and M. K. Belmonte. 2005. "Sex differences in the brain: Implications for explaining autism." *Science* 310 (5749): 819–23.

Baron-Cohen, Simon, Alan M. Leslie, and Uta Frith. 1985. "Does the autistic child have a 'theory of mind'?" *Cognition* 21 (1): 37–46.

Baron-Cohen, Simon, Fiona J. Scott, Carrie Allison, Joanna Williams, Patrick Bolton, Fiona E. Matthews, and Carol Brayne. 2009. "Prevalence of autism-spectrum conditions: UK school-based population study." *Brit J Psychiat* 194 (6): 500–509.

Baron-Cohen, S. and S. Wheelwright. 2004. "The empathy quotient: An investigation of adults with Asperger syndrome or high functioning autism, and normal sex differences." *J Autism Dev Disord* 34 (2): 163–75.

Baron-Cohen, Simon, Sally Wheelwright, Amy Burtenshaw, and Esther Hobson. 2007. "Mathematical talent is linked to autism." *Hum Nature* 18 (2): 125–31.

Baron-Cohen, Simon, Sally Wheelwright, Carol Stott, Patrick Bolton, and Ian Goodyer. 1997. "Is there a link between engineering and autism?" *Autism* 1:101–9.

Barttfeld, Pablo, Bruno Wicker, Sebastián Cukier, Silvana Navarta, Sergio Lew, Ramón Leiguarda, and Mariano Sigman. 2012. "State-dependent changes of connectivity patterns and functional brain network topology in autism spectrum disorder." *Neuropsychologia* 50 (14): 3653–62.

Baxter, Amanda J., T. S. Brugha, H. E. Erskine, R. W. Scheurer, Theo Vos, and J. G. Scott. 2015. "The epidemiology and global burden of autism spectrum disorders." *Psychol Med* 45 (3): 601–13.

Beacher, F. D., L. Minati, S. Baron-Cohen, M. V. Lombardo, M. C. Lai, M. A. Gray, N. A. Harrison, and H. D. Critchley. 2012. "Autism attenuates sex differences in brain structure: A combined voxel-based morphometry and diffusion tensor imaging study." *Am J Neuroradiol* 33 (1): 83–89.

Beacher, F. D., E. Radulescu, L. Minati, S. Baron-Cohen, M. V. Lombardo, M. C. Lai, A. Walker, D. Howard, M. A. Gray, N. A. Harrison, and H. D. Critchley. 2012. "Sex differences and autism: Brain function during verbal fluency and mental rotation." *PLoS One* 7 (6).

Beaudet, A. L. 2012. "Neuroscience: Preventable forms of autism?" *Science* 338 (6105): 342–43.

Bejerot, Susanne, and Jonna M. Eriksson. 2014. "Sexuality and gender role in autism spectrum disorder: A case control study." *PloS One* 9 (1).

Bejerot, Susanne, Jonna M. Eriksson, Sabina Bonde, Kjell Carlström, Mats B. Humble, and Elias Eriksson. 2012. "The extreme male brain revisited: Gender coherence in adults with autism spectrum disorder." *Brit J Psychiat* 201 (2): 116–23.

Bennett, T., P. Szatmari, S. Bryson, J. Volden, L. Zwaigenbaum, L. Vaccarella, E. Duku, and M. Boyle. 2008. "Differentiating autism and Asperger syndrome on the basis of language delay or impairment." *J Autism Dev Disord* 38 (4): 616–25.

Bettelheim, Bruno. 1967. *Infantile autism and the birth of the self.* New York: Free Press.

Bidinosti, Michael, Paolo Botta, Sebastian Krüttner, Catia C. Proenca, Natacha Stoehr, Mario Bernhard, Isabelle Fruh, Matthias Mueller, Debora Bonenfant, and Hans Voshol. 2016. "CLK2 inhibition ameliorates autistic features associated with SHANK3 deficiency." *Science* 351 (6278): 1199–1203.

Blakemore, S. J., D. Wolpert, and C. Frith. 2000. "Why can't you tickle yourself?" *Neuroreport* 11 (11): R11-6.

Bloemen, O. J., Q. Deeley, F. Sundram, E. M. Daly, G. J. Barker, D. K. Jones, T. A. van Amelsvoort, N. Schmitz, D. Robertson, K. C. Murphy, and D. G. Murphy. 2010. "White matter integrity in Asperger syndrome: A preliminary diffusion tensor magnetic resonance imaging study in adults." *Autism Res* 3 (5): 203–13.

Bloom, Paul. 2014. "Against empathy." *Boston Review.* http://bostonreview.net/forum/paul-bloom-against-empathy.

Blumberg, Stephen J., Matthew D. Bramlett, Michael D. Kogan, Laura A. Schieve, Jessica R. Jones, and Michael C. Lu. 2013. "Changes in prevalence of parent-reported autism spectrum disorder in school-aged US children: 2007 to 2011–2012." *Natl Health Stat Report* 65 (20): 1–7.

Bonnel, A., L. Mottron, I. Peretz, M. Trudel, E. Gallun, and A. M. Bonnel. 2003. "Enhanced pitch sensitivity in individuals with autism: A signal detection analysis." *J Cogn Neurosci* 15 (2): 226–35.

Bowlby, J. 1982. *Attachment and loss.* Vol. 1, 2nd ed. New York: Basic Books.

Brosnan, Mark, Rajiv Daggar, and John Collomosse. 2010. "The relationship between systemising and mental rotation and the implications for the extreme male brain theory of autism." *J Autism Dev Disord* 40 (1): 1–7.

Brown, David. 2006. "The forgotten sense—proprioception." *Dbl Review,* 20–24.

——. 2007. "The vestibular sense." *Dbl Review,* 17–22.

Broyd, S. J., C. Demanuele, S. Debener, S. K. Helps, C. J. James, and E. J. Sonuga-Barke. 2009. "Default-mode brain dysfunction in mental disorders: A systematic review." *Neurosci Biobehav Rev* 33 (3): 279–96.

Bucaille, Aurélie, Marine Grandgeorge, Céline Degrez, Camille Mallégol, Philippe Cam, Michel Botbol, and Pascale Planche. 2016. "Cognitive profile in adults with Asperger syndrome using WAIS-IV: Comparison to typical adults." *Res Autism Spect Dis* 21:1–9.

Buchen, L. 2011. "Scientists and autism: When geeks meet." *Nature* 479 (7371): 25–27.

Buckner, R. L., J. R. Andrews-Hanna, and D. L. Schacter. 2008. "The brain's default network: Anatomy, function, and relevance to disease." *Ann N Y Acad Sci* 1124:1–38.

Bullmore, Ed, and Olaf Sporns. 2012. "The economy of brain network organization." *Nat Rev Neurosci* 13 (5): 336–49.

Butler, T., J. Imperato-McGinley, H. Pan, D. Voyer, J. Cordero, Y. S. Zhu, E. Stern, and D. Silbersweig. 2006. "Sex differences in mental rotation: Top-down versus bottom-up processing." *Neuroimage* 32 (1): 445–56.

Buxhoeveden, D. P., K. Semendeferi, J. Buckwalter, N. Schenker, R. Switzer, and E. Courchesne. 2006. "Reduced minicolumns in the frontal cortex of patients with autism." *Neuropath Appl Neuro* 32 (5): 483–91.

Byrne, Richard W., and Andrew Whiten. 1988. *Machiavellian intelligence: Social expertise and the evolution of intellect in monkeys, apes, and humans.* Oxford: Oxford University Press.

Bzdok, D., R. Langner, L. Schilbach, O. Jakobs, C. Roski, S. Caspers, A. R. Laird, P. T. Fox, K. Zilles, and S. B. Eickhoff. 2013. "Characterization of the temporo-parietal junction by combining data-driven parcellation, complementary connectivity analyses, and functional decoding." *Neuroimage* 81:381–92.

Cahill, L. 2006. "Why sex matters for neuroscience." *Nat Rev Neurosci* 7 (6): 477–84.

Campbell, D. J., J. Chang, and K. Chawarska. 2014. "Early generalized overgrowth in autism spectrum disorder: Prevalence rates, gender effects, and clinical outcomes." *J Am Acad Child Adolesc Psychiatry* 53 (10): 1063–73e5.

Carrion-Castillo, Amaia, Barbara Franke, and Simon E. Fisher. 2013. "Molecular genetics of dyslexia: An overview." *Dyslexia* 19 (4): 214–40.

Casanova, Manuel F., Daniel P. Buxhoeveden, Andrew E. Switala, and Emil Roy. 2002. "Asperger's syndrome and cortical neuropathology." *J Child Neurol* 17 (2): 142–45.

Cascio, C. J., J. H. Foss-Feig, J. Heacock, K. B. Schauder, W. A. Loring, B. P. Rogers, J. R. Pryweller, C. R. Newsom, J. Cockhren, A. Cao, and S. Bolton. 2014. "Affective neural response to restricted interests in autism spectrum disorders." *J Child Psychol Psychiatry* 55 (2): 162–71.

Cederlund, M., and C. Gillberg. 2004. "One hundred males with Asperger syndrome: A clinical study of background and associated factors." *Dev Med Child Neurol* 46 (10): 652–60.

Chakrabarti, B., F. Dudbridge, L. Kent, S. Wheelwright, G. Hill-Cawthorne, C. Allison, S. Banerjee-Basu, and S. Baron-Cohen. 2009. "Genes related to sex steroids, neural growth, and social-emotional behavior are associated with autistic traits, empathy, and Asperger syndrome." *Autism Res* 2 (3): 157–77.

Chamak, Brigitte, and Béatrice Bonniau. 2013. "Autism and social movements in France: Exploring cross-cultural differences." In *Worlds of autism*, edited by Michael Orsini and Joyce Davidson, 239–57. Minneapolis: University of Minnesota Press.

Chamak, B., B. Bonniau, E. Jaunay, and D. Cohen. 2008. "What can we learn about autism from autistic persons?" *Psychother Psychosom* 77 (5): 271–79.

Charlop, M. H., and L. K. Haymes. 1994. "Speech and language acquisition and intervention: Behavioral approaches." In *Autism in children and adults: Etiology, assessment, and intervention*, edited by J. L. Matson, 213–40. Pacific Grove, Calif.: Brooks/Cole.

Chaste, Pauline, Lambertus Klei, Stephan J. Sanders, Vanessa Hus, Michael T. Murtha, Jennifer K. Lowe, A. Jeremy Willsey, Daniel Moreno-De-Luca, W. Yu Timothy, and Eric Fombonne. 2015. "A genome-wide association study of autism using the Simons Simplex Collection: Does reducing phenotypic heterogeneity in autism increase genetic homogeneity?" *Biol psychiat* 77 (9): 775–84.

Chiang, H. M., L. Y. Tsai, Y. K. Cheung, A. Brown, and H. Li. 2014. "A meta-analysis of differences in IQ profiles between individuals with Asperger's disorder and high-functioning autism." *J Autism Dev Disord* 44 (7): 1577–96.

Christensen, D. L., J. Baio, K. Van Naarden Braun, D. Bilder, J. Charles, J. N. Constantino, J. Daniels, M. S. Durkin, R. T. Fitzgerald, M. Kurzius-Spencer,

L. C. Lee, S. Pettygrove, C. Robinson, E. Schulz, C. Wells, M. S. Wingate, W. Zahorodny, M. Yeargin-Allsopp, Control Centers for Disease, and Prevention. 2016. "Prevalence and characteristics of autism spectrum disorder among children aged 8 years—autism and developmental disabilities monitoring network, 11 sites, United States, 2012." *MMWR Surveill Summ* 65 (3): 1–23.

Cohen, D., R. S. Cassel, C. Saint-Georges, A. Mahdhaoui, M. C. Laznik, F. Apicella, P. Muratori, S. Maestro, F. Muratori, and M. Chetouani. 2013. "Do parentese prosody and fathers' involvement in interacting facilitate social interaction in infants who later develop autism?" *PLoS One* 8 (5): e61402.

Colvert, Emma, Beata Tick, Fiona McEwen, Catherine Stewart, Sarah R. Curran, Emma Woodhouse, Nicola Gillan, Victoria Hallett, Stephanie Lietz, and Tracy Garnett. 2015. "Heritability of autism spectrum disorder in a UK population-based twin sample." *JAMA psychiatry* 72 (5): 415–23.

Conson, Massimiliano, Elisabetta Mazzarella, Alessandro Frolli, Dalila Esposito, Nicoletta Marino, Luigi Trojano, Angelo Massagli, Giovanna Gison, Nellantonio Aprea, and Dario Grossi. 2013. "Motor imagery in Asperger syndrome: testing action simulation by the hand laterality task." *PLoS One* 8 (7): e70734.

Coren, S., C. Porac, and L. M. Ward. 1984. *Sensation and perception.* 2nd ed. New York: Academic Press.

Cosgrove, Kelly P., Carolyn M. Mazure, and Julie K. Staley. 2007. "Evolving knowledge of sex differences in brain structure, function, and chemistry." *Biol psychiat* 62 (8): 847–55.

Costanzo, E. Y., M. Villarreal, L. J. Drucaroff, M. Ortiz-Villafane, M. N. Castro, M. Goldschmidt, A. E. Wainsztein, M. S. Ladron-de-Guevara, C. Romero, L. I. Brusco, J. A. Camprodon, C. Nemeroff, and S. M. Guinjoan. 2015. "Hemispheric specialization in affective responses, cerebral dominance for language, and handedness: Lateralization of emotion, language, and dexterity." *Behav Brain Res* 288:11–19.

Courchesne, E., K. Campbell, and S. Solso. 2011. "Brain growth across the life span in autism: Age-specific changes in anatomical pathology." *Brain Res* 1380:138–45.

Courchesne, E., H. J. Chisum, J. Townsend, A. Cowles, J. Covington, B. Egaas, M. Harwood, S. Hinds, and G. A. Press. 2000. "Normal brain development and aging: Quantitative analysis at in vivo MR imaging in healthy volunteers." *Radiology* 216 (3): 672–82.

Courchesne, Eric, Peter Mouton, Michael E. Calhoun, Katerina Semendeferi, Clelia Ahrens-Barbeau, Melodie J. Hallet, Cynthia Carter Barnes, and Karen Pierce. 2011. "Neuron number and size in prefrontal cortex of children with autism." *JAMA* 306 (18): 2001–10.

Crespi, Bernard. 2006. "The natural selection of psychosis." *Behav Brain Sci* 29 (04): 410–11.

———. 2013. "Developmental heterochrony and the evolution of autistic perception, cognition and behavior." *BMC Med* 11 (1): 119.

Crowley, David, and Paul Heyer. 2016. *Communication in history: Technology, culture, society.* 6th ed. New York: Routledge.

Csikszentmihalyi, Mihaly. 1990. *Flow: The psychology of optimal performance.* New York: Cambridge University Press.

Dahlin, Kristina, Margaret Taylor, and Mark Fichman. 2004. "Today's Edisons or weekend hobbyists: Technical merit and success of inventions by independent inventors." *Res Policy* 33 (8): 1167–83.

Daley, Tamara C. 2002. "The need for cross-cultural research on the pervasive developmental disorders." *Transcult Psychiatry* 39 (4): 531–50.

Daley, Tamara C., Nidhi Singhal, and Vibha Krishnamurthy. 2013. "Ethical considerations in conducting research on autism spectrum disorders in low and middle income countries." *J Autism Dev Disord* 43 (9): 2002–14.

Dalton, Nicholas Sheep. 2013. "Neurodiversity & HCI." *interactions* 20, no. 2 (March/April 2013): 72–75.

Damoiseaux, J. S., S. A. R. B. Rombouts, F. Barkhof, P. Scheltens, C. J. Stam, Stephen M. Smith, and C. F. Beckmann. 2006. "Consistent resting-state networks across healthy subjects." *Proc Natl Acad Sci USA* 103 (37): 13848–53.

Dane, S., and N. Balci. 2007. "Handedness, eyedness and nasal cycle in children with autism." *Int J Dev Neurosci* 25 (4): 223–26.

Dapretto, M., M. S. Davies, J. H. Pfeifer, A. A. Scott, M. Sigman, S. Y. Bookheimer, and M. Iacoboni. 2006. "Understanding emotions in others: Mirror neuron dysfunction in children with autism spectrum disorders." *Nat Neurosci* 9 (1): 28–30.

Darwin, Charles. 1859. *On the origin of species by means of natural selection.* London: J. Murray.

———. 1871. *The Descent of Man, and Selection in Relation to Sex.* New York: D. Appleton.

Dave, Dhaval M., and Jose M. Fernandez. 2015. "Rising autism prevalence: Real or displacing other mental disorders? Evidence from demand for auxiliary healthcare workers in California." *Econ Inq* 53 (1): 448–68.

Davidson, Joyce. 2008. "Autistic culture online: Virtual communication and cultural expression on the spectrum." *Soc Cult Geogr* 9 (7): 791–806.

Day, Jeremy J., and J. David Sweatt. 2011. "Epigenetic mechanisms in cognition." *Neuron* 70 (5): 813–29.

Dehaene, Stanislas, and Laurent Cohen. 2007. "Cultural recycling of cortical maps." *Neuron* 56 (2): 384–98.

Dehaene, Stanislas, Laurent Cohen, José Morais, and Régine Kolinsky. 2015. "Illiterate to literate: Behavioural and cerebral changes induced by reading acquisition." *Nat Rev Neurosci* 16 (4): 234–44.

Demaree, H. A., D. E. Everhart, E. A. Youngstrom, and D. W. Harrison. 2005. "Brain lateralization of emotional processing: historical roots and a future incorporating 'dominance.'" *Behav Cogn Neurosci Rev* 4 (1): 3–20.

Dementieva, Y. A., D. D. Vance, S. L. Donnelly, L. A. Elston, C. M. Wolpert, S. A. Ravan, G. R. DeLong, R. K. Abramson, H. H. Wright, and M. L. Cuccaro. 2005. "Accelerated head growth in early development of individuals with autism." *Pediatr Neurol* 32 (2): 102–8.

DeSilva, J. M. 2016. "Brains, birth, bipedalism and the mosaic evolution of the helpless human infant." In *Costly and cute: How helpless newborns made us human*, edited by W. Trevathan and K. Rosenberg, 67–86. Santa Fe, N.M.: SAR Press; Albuquerque: University of New Mexico Press.

Des Roches Rosa, Shannon. 2016. "Before talking about autism, listen to families." *Spectrum News*, January 26. https://spectrumnews.org/opinion/viewpoint/before-talking-about-autism-listen-to-families.

Developmental, Disabilities Monitoring Network Surveillance Year, and Principal Investigators. 2014. "Prevalence of autism spectrum disorder among children aged 8 years—autism and developmental disabilities monitoring network, 11 sites, United States, 2010." *MMWR Surveill Summ* 63 (2): 1.

De Waal, Frans B. M. 2008. "Putting the altruism back into altruism: The evolution of empathy." *Annu Rev Psychol* 59:279–300.

Dhiman, Satinder. 2012. "Mindfulness and the art of living creatively: Cultivating a creative life by minding our mind." *J Soc Change* 4 (1): 1.

Di Martino, Adriana, Kathryn Ross, Lucina Q. Uddin, Andrew B. Sklar, F. Xavier Castellanos, and Michael P. Milham. 2009. "Functional brain correlates of social and nonsocial processes in autism spectrum disorders: An activation likelihood estimation meta-analysis." *Biol Psychiat* 65 (1): 63–74.

Di Martino, A., C. G. Yan, Q. Li, E. Denio, F. X. Castellanos, K. Alaerts, J. S. Anderson, M. Assaf, S. Y. Bookheimer, M. Dapretto, B. Deen, S. Delmonte, I. Dinstein, B. Ertl-Wagner, D. A. Fair, L. Gallagher, D. P. Kennedy, C. L. Keown, C. Keysers, J. E. Lainhart, C. Lord, B. Luna, V. Menon, N. J. Minshew, C. S. Monk, S. Mueller, R. A. Muller, M. B. Nebel, J. T. Nigg, K. O'Hearn, K. A. Pelphrey, S. J. Peltier, J. D. Rudie, S. Sunaert, M. Thioux, J. M. Tyszka, L. Q. Uddin, J. S. Verhoeven, N. Wenderoth, J. L. Wiggins, S. H. Mostofsky, and M. P. Milham. 2014. "The autism brain imaging data exchange: Towards a large-scale evaluation of the intrinsic brain architecture in autism." *Mol Psychiatry* 19 (6): 659–67.

Di Napoli, A., V. Warrier, S. Baron-Cohen, and B. Chakrabarti. 2014. "Genetic variation in the oxytocin receptor (*OXTR*) gene is associated with Asperger syndrome." *Mol Autism* 5 (1): 48. doi:10.1186/2040-2392-5-48.

———. 2015. "Genetic variant rs17225178 in the *ARNT2* gene is associated with Asperger syndrome." *Mol Autism* 6 (1): 9.

Dinstein, I., K. Pierce, L. Eyler, S. Solso, R. Malach, M. Behrmann, and E. Courchesne. 2011. "Disrupted neural synchronization in toddlers with autism." *Neuron* 70 (6): 1218–25.

Dunbar, Robin I. M., and Susanne Shultz. 2007. "Evolution in the social brain." *Science* 317 (5843): 1344–47.

Durdiakova, J., V. Warrier, S. Banerjee-Basu, S. Baron-Cohen, and B. Chakrabarti.

2014. "STX1A and Asperger syndrome: A replication study." *Mol Autism* 5 (1): 14.

Durdiakova, J., V. Warrier, S. Baron-Cohen, and B. Chakrabarti. 2014. "Single nucleotide polymorphism rs6716901 in SLC25A12 gene is associated with Asperger syndrome." *Mol Autism* 5 (1): 25.

Dvash, Jonathan, and Simone G. Shamay-Tsoory. 2014. "Theory of mind and empathy as multidimensional constructs: Neurological foundations." *Top Lang Disord* 34 (4): 282–95.

Ebisch, S. J., V. Gallese, R. M. Willems, D. Mantini, W. B. Groen, G. L. Romani, J. K. Buitelaar, and H. Bekkering. 2011. "Altered intrinsic functional connectivity of anterior and posterior insula regions in high-functioning participants with autism spectrum disorder." *Hum Brain Mapp* 32 (7): 1013–28.

Ecker, C., S. Y. Bookheimer, and D. G. Murphy. 2015. "Neuroimaging in autism spectrum disorder: Brain structure and function across the lifespan." *Lancet Neurol* 14 (11): 1121–34.

Eibl-Eibesfeldt, Irenäus. 1989. *Human ethology.* New York: Aldine de Gruyter.

El-Fishawy, Paul, and Matthew W. State. 2010. "The genetics of autism: Key issues, recent findings, and clinical implications." *Psychiatr Clin North Am* 33 (1): 83–105.

Ellis, Hadyn D., Diane M. Ellis, William Fraser, and Shoumitro Deb. 1994. "A preliminary study of right hemisphere cognitive deficits and impaired social judgments among young people with Asperger syndrome." *Eur Child Adolesc Psychiatry* 3 (4): 255–66.

Ellis, Lee. 2011. "Identifying and explaining apparent universal sex differences in cognition and behavior." *Pers Indiv Differ* 51 (5): 552–61.

Elsabbagh, Mayada, Gauri Divan, Yun-Joo Koh, Young Shin Kim, Shuaib Kauchali, Carlos Marcín, Cecilia Montiel-Nava, Vikram Patel, Cristiane S. Paula, and Chongying Wang. 2012. "Global prevalence of autism and other pervasive developmental disorders." *Autism Res* 5 (3): 160–79.

Else-Quest, Nicole M., Janet Shibley Hyde, and Marcia C. Linn. 2010. "Cross-national patterns of gender differences in mathematics: A meta-analysis." *Psychol Bull* 136 (1): 103.

Elton, A., and W. Gao. 2015. "Task-positive functional connectivity of the default mode network transcends task domain." *J Cogn Neurosci,* 1–13.

Ernsperger, Lori, and Danielle Wendel. 2007. *Girls under the umbrella of autism spectrum disorders: Practical solutions for addressing everyday challenges:* Lenexa, Kans.: AAPC.

Escalante-Mead, Paul R., Nancy J. Minshew, and John A. Sweeney. 2003. "Abnormal brain lateralization in high-functioning autism." *J Autism Dev Disord* 33 (5): 539–43.

Evans, Sarah, Nick Neave, and Delia Wakelin. 2006. "Relationships between vocal characteristics and body size and shape in human males: An evolutionary explanation for a deep male voice." *Biol Psychol* 72 (2): 160–63.

Falk, D. 2004. "Prelinguistic evolution in early hominins: Whence motherese?" *Behav Brain Sci* 27 (4): 491–541.

———. 2009. *Finding our tongues: Mothers, infants and the origins of language*. New York: Pertheus / Basic Books.

———. 2011. *The Fossil chronicles: How two controversial discoveries changed our view of human evolution*. Berkeley and Los Angeles: University of California Press.

———. 2012. "The adaptive value of happiness: Evolved trait or evolutionary useless spinoff?" *Being human*. http://www.beinghuman.org/big-questions/article/adaptive-value-happiness.

———. 2012. "Charles Darwin, the first paleoanthropologist." *Hum Evol* 7 (4): 217–29.

———. 2016. "Baby the trendsetter: Three evo-devo trends and their expression in Asperger syndrome." In *Costly and cute: How helpless newborns made us human*, edited by W. Trevathan and K. Rosenberg, 109–132. Santa Fe, N.M.: SAR Press; Albuquerque: University of New Mexico Press.

———. 2016. "Evolution of brain and culture: The neurological and cognitive journey from *Australopithecus* to Albert Einstein." *J Anthropol Sci* 94:1–14.

Falk, D., N. Froese, D. S. Sade, and B. C. Dudek. 1999. "Sex differences in brain/body relationships of rhesus monkeys and humans." *J Hum Evol* 36 (2): 233–38.

Falter, C. M., K. C. Plaisted, and G. Davis. 2008. "Visuo-spatial processing in autism—testing the predictions of extreme male brain theory." *J Autism Dev Disord* 38 (3): 507–15.

Farb, Norman A. S., Zindel V. Segal, Helen Mayberg, Jim Bean, Deborah McKeon, Zainab Fatima, and Adam K. Anderson. 2007. "Attending to the present: Mindfulness meditation reveals distinct neural modes of self-reference." *Soc Cogn Affect Neurosci* 2 (4): 313–22.

Feldman, Ruth, Mikhail Monakhov, Maayan Pratt, and Richard P. Ebstein. 2016. "Oxytocin pathway genes: Evolutionary ancient system impacting on human affiliation, sociality, and psychopathology." *Biol Psychiat* 79 (3): 174–84.

Fernald, Anne. 1994. "Human maternal vocalizations to infants as biologically relevant signals: An evolutionary perspective." In *Language acquisition core readings*, edited by P. Bloom, Cambridge, 51–94. Cambridge, Mass.: MIT Press.

Fine, Cordelia. 2014. "His brain, her brain?" *Science* 346 (6212): 915–16.

Fiser, J., and R. N. Aslin. 2002. "Statistical learning of new visual feature combinations by infants." *Proc Natl Acad Sci U S A* 99 (24): 15822–26.

Floris, Dorothea L., Meng-Chuan Lai, Tibor Auer, Michael V. Lombardo, Christine Ecker, Bhismadev Chakrabarti, Sally J. Wheelwright, Edward T. Bullmore, Declan G. M. Murphy, and Simon Baron-Cohen. 2016. "Atypically rightward cerebral asymmetry in male adults with autism stratifies individuals with and without language delay." *Hum Brain Mapp* 37 (1): 230–53.

Foley-Nicpon, M., S. G. Assouline, and R. D. Stinson. 2012. "Cognitive and academic

distinctions between gifted students with autism and Asperger syndrome."
Gifted Child Quart 56 (2): 77–89.

Fombonne, Eric. 2009. "Epidemiology of pervasive developmental disorders." *Pediatr Res* 65 (6): 591–98.

Fox, M. D., A. Z. Snyder, J. L. Vincent, M. Corbetta, D. C. Van Essen, and M. E. Raichle. 2005. "The human brain is intrinsically organized into dynamic, anticorrelated functional networks." *Proc Natl Acad Sci U S A* 102 (27): 9673–78.

Frazier, Thomas W., Stelios Georgiades, Somer L. Bishop, and Antonio Y. Hardan. 2014. "Behavioral and cognitive characteristics of females and males with autism in the Simons Simplex Collection." *J Am Acad Child Adolesc Psychiatry* 53 (3): 329–40.

Frith, Uta. 2004. "Emanuel Miller lecture: Confusions and controversies about Asperger syndrome." *J Child Psychol Psychiatry* 45 (4): 672–86.

Futagi, Y., and Y. Suzuki. 2010. "Neural mechanism and clinical significance of the plantar grasp reflex in infants." *Pediatr Neurol* 43 (2): 81–86.

Gaugler, Trent, Lambertus Klei, Stephan J. Sanders, Corneliu A. Bodea, Arthur P. Goldberg, Ann B. Lee, Milind Mahajan, Dina Manaa, Yudi Pawitan, and Jennifer Reichert. 2014. "Most genetic risk for autism resides with common variation." *Nat Genet* 46 (8): 881–85.

Gauthier, Julie, and Guy A. Rouleau. 2011. "A new genetic mechanism for autism." In *Autism spectrum disorders: The role of genetics in diagnosis and treatment*, edited by S. I. Deutsch and M. R. Urbano, 103–24. Rijeka, Croatia: InTech.

Gazzaniga, M. S. 2005. "Forty-five years of split-brain research and still going strong." *Nat Rev Neurosci* 6 (8): 653–59.

Geier, David A., Janet K. Kern, Paul G. King, Lisa K. Sykes, and Mark R. Geier. 2012. "An evaluation of the role and treatment of elevated male hormones in autism spectrum disorders." *Acta Neurobiol Exp (Wars)* 72 (1): 1–17.

Gernsbacher, Morton Ann, Michelle Dawson, and Laurent Mottron. 2006. "Autism: Common, heritable, but not harmful." *Behav Brain Sci* 29 (4): 413–14.

Geschwind, D. H. 2011. "Genetics of autism spectrum disorders." *Trends Cogn Sci* 15 (9): 409–16.

Ghaziuddin, M. 2008. "Defining the behavioral phenotype of Asperger syndrome." *J Autism Dev Disord* 38 (1): 138–42.

———. 2010. "Should the DSM V drop Asperger syndrome?" *J Autism Dev Disord* 40 (9): 1146–48.

Ghaziuddin, Mohammad, and Kimberly Mountain-Kimchi. 2004. "Defining the intellectual profile of Asperger syndrome: Comparison with high-functioning autism." *J Autism Dev Disord* 34 (3): 279–84.

Gibbons, Ann. 2016. "Five matings for moderns, Neandertals." *Science* 351 (6279): 1250–51.

Gilbert, D. T., and T. D. Wilson. 2007. "Prospection: Experiencing the future." *Science* 317 (5843): 1351–54.

Gillberg, Christopher. 2002. *A guide to Asperger syndrome*. New York: Cambridge University Press.

Gillberg, I. C., and C. Gillberg. 1989. "Asperger syndrome—some epidemiological considerations: A research note." *J Child Psychol Psychiatry* 30 (4): 631–38.

Gillberg, C., and L. de Souza. 2002. "Head circumference in autism, Asperger syndrome, and ADHD: A comparative study." *Dev Med Child Neurol* 44 (5): 296–300.

Gilmore, J. H., W. Lin, I. Corouge, Y. S. K. Vetsa, J. Keith Smith, C. Kang, H. Gu, R. M. Hamer, J. A. Lieberman, and G. Gerig. 2007. "Early postnatal development of corpus callosum and corticospinal white matter assessed with quantitative tractography." *Am J Neuroradiol* 28 (9): 1789–95.

Gilmore, J. H., W. Lin, M. W. Prastawa, C. B. Looney, Y. S. Vetsa, R. C. Knickmeyer, D. D. Evans, J. K. Smith, R. M. Hamer, J. A. Lieberman, and G. Gerig. 2007. "Regional gray matter growth, sexual dimorphism, and cerebral asymmetry in the neonatal brain." *J Neurosci* 27 (6): 1255–60.

Goin-Kochel, Robin P., Anna Abbacchi, John N. Constantino, and Autism Genetic Resource Exchange Consortium. 2007. "Lack of evidence for increased genetic loading for autism among families of affected females: A replication from family history data in two large samples." *Autism* 11 (3): 279–86.

Gold, R., and M. Faust. 2010. "Right hemisphere dysfunction and metaphor comprehension in young adults with Asperger syndrome." *J Autism Dev Disord* 40 (7): 800–811.

Goldberger, Z. D. 2001. "Music of the left hemisphere: Exploring the neurobiology of absolute pitch." *Yale J Biol Med* 74 (5): 323–27.

Golden, Hays. 2013. "Childhood autism and assortative mating." PhD diss., University of Chicago.

Gong, G., P. Rosa-Neto, F. Carbonell, Z. J. Chen, Y. He, and A. C. Evans. 2009. "Age- and gender-related differences in the cortical anatomical network." *J Neurosci* 29 (50): 15684–93.

Gotts, Stephen J., Hang Joon Jo, Gregory L. Wallace, Ziad S. Saad, Robert W. Cox, and Alex Martin. 2013. "Two distinct forms of functional lateralization in the human brain." *Proc Natl Acad Sci U S A* 110 (36): E3435–44.

Grandin, T. 1995. *Thinking in pictures and other reports from my life with autism*. New York: Vintage Books.

Grandin, Temple, and Margaret M. Scariano. 1986. *A true story: Emergence, labeled autistic*. New York: Warner Books.

Green, C. Shawn, and Daphne Bavelier. 2003. "Action video game modifies visual selective attention." *Nature* 423 (6939): 534–37.

Green, Cherie, Cheryl Dissanayake, and Danuta Loesch. 2015. "A review of physical growth in children and adolescents with autism spectrum disorder." *Dev Rev* 36:156–78.

Green, R. E., J. Krause, A. W. Briggs, T. Maricic, U. Stenzel, M. Kircher, N. Patterson, H. Li, W. Zhai, M. H. Fritz, N. F. Hansen, E. Y. Durand, A. S. Malaspinas,

J. D. Jensen, T. Marques-Bonet, C. Alkan, K. Prufer, M. Meyer, H. A.
Burbano, J. M. Good, R. Schultz, A. Aximu-Petri, A. Butthof, B. Hober,
B. Hoffner, M. Siegemund, A. Weihmann, C. Nusbaum, E. S. Lander,
C. Russ, N. Novod, J. Affourtit, M. Egholm, C. Verna, P. Rudan, D. Bra-
jkovic, Z. Kucan, I. Gusic, V. B. Doronichev, L. V. Golovanova, C. Lalueza-
Fox, M. de la Rasilla, J. Fortea, A. Rosas, R. W. Schmitz, P. L. Johnson, E. E.
Eichler, D. Falush, E. Birney, J. C. Mullikin, M. Slatkin, R. Nielsen, J. Kelso,
M. Lachmann, D. Reich, and S. Paabo. 2010. "A draft sequence of the Nean-
dertal genome." *Science* 328 (5979): 710–22.

Green, Shulamite A., Jeffrey D. Rudie, Natalie L. Colich, Jeffrey J. Wood, David
Shirinyan, Leanna Hernandez, Nim Tottenham, Mirella Dapretto, and
Susan Y. Bookheimer. 2013. "Overreactive brain responses to sensory stim-
uli in youth with autism spectrum disorders." *J Am Acad Child Adolesc
Psychiatry* 52 (11): 1158–72.

Grigorenko, Elena L., Ami Klin, and Fred Volkmar. 2003. "Annotation: Hyperlexia:
Disability or superability?" *J Child Psychol Psychiatry* 44 (8): 1079–91.

Grinker, Roy Richard. 2008. *Unstrange minds: Remapping the world of autism*. Cambridge, Mass.: Da Capo Press.

Grinker, R. R., N. Chambers, N. Njongwe, A. E. Lagman, W. Guthrie, S. Stronach,
B. O. Richard, S. Kauchali, B. Killian, M. Chhagan, F. Yucel, M. Kudumu,
C. Barker-Cummings, J. Grether, and A. M. Wetherby. 2012. "'Communi-
ties' in community engagement: Lessons learned from autism research in
South Korea and South Africa." *Autism Res* 5 (3): 201–10.

Grodzinsky, Y., and I. Nelken. 2014. "Neuroscience: The neural code that makes us
human." *Science* 343 (6174): 978–79.

Gunter, Chris. 2015. "Autism: Seeing the spectrum entire." *Nature* 524 (7565): 288–89.

Gunter, H. L., M. Ghaziuddin, and H. D. Ellis. 2002. "Asperger syndrome: Tests of
right hemisphere functioning and interhemispheric communication."
J Autism Dev Disord 32 (4): 263–81.

Hall, Kenneth. 2001. *Asperger syndrome, the universe and everything*. Philadelphia:
Jessica Kingsley.

Hallmayer, J., S. Cleveland, A. Torres, J. Phillips, B. Cohen, T. Torigoe, J. Miller, A. Fed-
ele, J. Collins, K. Smith, L. Lotspeich, L. A. Croen, S. Ozonoff, C. Lajonchere,
J. K. Grether, and N. Risch. 2011. "Genetic heritability and shared environmental factors among twin pairs with autism." *Arch Gen Psychiatry* 68 (11):
1095–102.

Hansen, S. N., D. E. Schendel, and E. T. Parner. 2015. "Explaining the increase in the
prevalence of autism spectrum disorders: The proportion attributable to
changes in reporting practices." *JAMA Pediatr* 169 (1): 56–62.

Happé, F., and U. Frith. 2006. "The weak coherence account: detail-focused cognitive
style in autism spectrum disorders." *J Autism Dev Disord* 36 (1): 5–25.

Happe, F., and P. Vital. 2009. "What aspects of autism predispose to talent?" *Philos
Trans R Soc Lond B Biol Sci* 364 (1522): 1369–75.

Hardan, Antonio Y., Ragy R. Girgis, Jason Adams, Andrew R. Gilbert, Nadine M. Melhem, Matcheri S. Keshavan, and Nancy J. Minshew. 2008. "Brief report: Abnormal association between the thalamus and brain size in Asperger's disorder." *J Autism Dev Disord* 38 (2): 390–94.

Harlow, Harry F. 1958. "The nature of love." *Am Psychol* 13:573–685.

Harmand, S., J. E. Lewis, C. S. Feibel, C. J. Lepre, S. Prat, A. Lenoble, X. Boes, R. L. Quinn, M. Brenet, A. Arroyo, N. Taylor, S. Clement, G. Daver, J. P. Brugal, L. Leakey, R. A. Mortlock, J. D. Wright, S. Lokorodi, C. Kirwa, D. V. Kent, and H. Roche. 2015. "3.3-million-year-old stone tools from Lomekwi 3, West Turkana, Kenya." *Nature* 521 (7552): 310–15.

Hawks, J. 2013. "Archaic human genomes and language evolution." *J Anthropol Sci* 91:253–55.

Hazlett, H. C., H. Gu, B. C. Munsell, S. H. Kim, M. Styner, J. J. Wolff, J. T. Elison, M. R. Swanson, H. Zhu, K. N. Botteron, D. L. Collins, J. N. Constantino, S. R. Dager, A. M. Estes, A. C. Evans, V. S. Fonov, G. Gerig, P. Kostopoulos, R. C. McKinstry, J. Pandey, S. Paterson, J. R. Pruett, R. T. Schultz, D. W. Shaw, L. Zwaigenbaum, J. Piven, and IBIS Network. 2017. "Early brain development in infants at high risk for autism spectrum disorder." *Nature* 542 (7641): 348–51.

Henrich, Joseph, Steven J. Heine, and Ara Norenzayan. 2010. "The weirdest people in the world?" *Behav Brain Sci* 33 (2–3): 61–83.

Hermoye, Laurent, Christine Saint-Martin, Guy Cosnard, Seung-Koo Lee, Jinna Kim, Marie-Cecile Nassogne, Renaud Menten, Philippe Clapuyt, Pamela K. Donohue, and Kegang Hua. 2006. "Pediatric diffusion tensor imaging: normal database and observation of the white matter maturation in early childhood." *Neuroimage* 29 (2): 493–504.

Hill, Alison Presmanes, Katharine Zuckerman, and Eric Fombonne. 2015. "Epidemiology of autism spectrum disorders." In *Translational approaches to autism spectrum disorder*, edited by M. Robinson-Agramonte, 13–38. New York: Springer.

Hines, M. 2011. "Prenatal endocrine influences on sexual orientation and on sexually differentiated childhood behavior." *Front Neuroendocrinol* 32 (2): 170–82.

Hofman, Michel A. 2014. "Evolution of the human brain: When bigger is better." *Front Neuroanat* 8:15. doi:10.3389/fnana.2014.00015.

Hrdy, S. 2016. "Of marmosets, men, and the transformative power of babies." In *Costly and cute: How helpless newborns made us human*, edited by W. Trevathan and K. Rosenberg, 177–203. Santa Fe, N.M.: SAR Press; Albuquerque: University of New Mexico Press.

Hugdahl, Kenneth, Tormod Thomsen, and Lars Ersland. 2006. "Sex differences in visuo-spatial processing: An fMRI study of mental rotation." *Neuropsychologia* 44 (9): 1575–83.

Hurlburt, R. T., F. Happé, and U. Frith. 1994. "Sampling the form of inner experience in three adults with Asperger syndrome." *Psychol Med* 24 (2): 385–95.

Hyde, Janet Shibley. 2005. "The gender similarities hypothesis." *Am Psychol* 60 (6): 581.

Iacoboni, Marco. 2009. "Imitation, empathy, and mirror neurons." *Annu Rev Psychol* 60:653–70.

Ingalhalikar, M., A. Smith, D. Parker, T. D. Satterthwaite, M. A. Elliott, K. Ruparel, H. Hakonarson, R. E. Gur, R. C. Gur, and R. Verma. 2014. "Sex differences in the structural connectome of the human brain." *Proc Natl Acad Sci U S A* 111 (2): 823–28.

Ingudomnukul, Erin, Simon Baron-Cohen, Sally Wheelwright, and Rebecca Knickmeyer. 2007. "Elevated rates of testosterone-related disorders in women with autism spectrum conditions." *Horm Behav* 51 (5): 597–604.

Iossifov, Ivan, Dan Levy, Jeremy Allen, Kenny Ye, Michael Ronemus, Yoon-ha Lee, Boris Yamrom, and Michael Wigler. 2015. "Low load for disruptive mutations in autism genes and their biased transmission." *Proc Natl Acad Sci U S A* 112 (41): E5600–607.

Iourov, Ivan Y., Svetlana G. Vorsanova, Victoria Y. Voinova, and Yuri B. Yurov. 2015. "3p22. 1p21. 31 microdeletion identifies *CCK* as Asperger syndrome candidate gene and shows the way for therapeutic strategies in chromosome imbalances." *Mol Cytogenet* 8 (1): 1.

Jaarsma, P., and S. Welin. 2012. "Autism as a natural human variation: Reflections on the claims of the neurodiversity movement." *Health Care Anal* 20 (1): 20–30.

Jackson, Luke. 2002. *Freaks, geeks and Asperger syndrome*. Philadelphia: Jessica Kingsley.

Joel, Daphna, Zohar Berman, Ido Tavor, Nadav Wexler, Olga Gaber, Yaniv Stein, Nisan Shefi, Jared Pool, Sebastian Urchs, and Daniel S. Margulies. 2015. "Sex beyond the genitalia: The human brain mosaic." *Proc Natl Acad Sci U S A* 112 (50): 15468–73.

Joseph, R. 2000. "The evolution of sex differences in language, sexuality, and visual-spatial skills." *Arch Sex Behav* 29 (1): 35–66.

Joshi, P. K., et al. (ROHgen consortium). 2015. "Directional dominance on stature and cognition in diverse human populations." *Nature* 523 (7561): 459–62.

Just, M. A., V. L. Cherkassky, T. A. Keller, R. K. Kana, and N. J. Minshew. 2007. "Functional and anatomical cortical underconnectivity in autism: evidence from an fMRI study of an executive function task and corpus callosum morphometry." *Cereb Cortex* 17 (4): 951–61.

Kaiser, Martha D., Daniel Y.-J. Yang, Avery C. Voos, Randi H. Bennett, Ilanit Gordon, Charlotte Pretzsch, Danielle Beam, Cara Keifer, Jeffrey Eilbott, and Francis McGlone. 2015. "Brain mechanisms for processing affective (and nonaffective) touch are atypical in autism." *Cereb Cortex*. doi:10.1093/cercor/bhv125.

Kaland, Nils. 2011. "Brief report: Should Asperger syndrome be excluded from the forthcoming DSM-V?" *Res Autism Spect Dis* 5 (3): 984–89.

Kaland, N., A. Moller-Nielsen, K. Callesen, E. L. Mortensen, D. Gottlieb, and L. Smith. 2002. "A new 'advanced' test of theory of mind: Evidence from children and adolescents with Asperger syndrome." *J Child Psychol Psychiatry* 43 (4): 517–28.

Kanaan, Richard A., Matthew Allin, Marco Picchioni, Gareth J. Barker, Eileen Daly, Sukhwinder S. Shergill, James Woolley, and Philip K. McGuire. 2012. "Gender differences in white matter microstructure." *PLoS One* 7 (6): e38272.

Kanherkar, Riya R., Naina Bhatia-Dey, and Antonei B. Csoka. 2014. "Epigenetics across the human lifespan." *Front Cell Dev Biol* 2 (49): 25364756. doi:10.3389/fcell.2014.00049.

Kanner, L. 1943. "Autistic disturbances of affective contact." *Nervous Child* 2:217–250.

———. 1944. "Early infantile autism." *J Pediatr* 25 (3): 211–17.

Karmiloff, Kyra, and Annette Karmiloff-Smith. 2001. *Pathways to language: From fetus to adolescent, the developing child.* Cambridge: Harvard University Press.

Katagiri, Masatoshi, Tetsuko Kasai, Yoko Kamio, and Harumitsu Murohashi. 2013. "Individuals with Asperger's disorder exhibit difficulty in switching attention from a local level to a global level." *J Autism Dev Disord* 43 (2): 395–403.

Keller, Matthew C., and Geoffrey Miller. 2006. "Resolving the paradox of common, harmful, heritable mental disorders: Which evolutionary genetic models work best?" *Behav Brain Sci* 29 (4): 385–404.

Kennedy, D. P., and E. Courchesne. 2008. "Functional abnormalities of the default network during self- and other-reflection in autism." *Soc Cogn Affect Neurosci* 3 (2): 177–90.

———. 2008. "The intrinsic functional organization of the brain is altered in autism." *Neuroimage* 39 (4): 1877–85.

Keyes, K. M., E. Susser, K. Cheslack-Postava, C. Fountain, K. Liu, and P. S. Bearman. 2012. "Cohort effects explain the increase in autism diagnosis among children born from 1992 to 2003 in California." *Int J Epidemiol* 41 (2): 495–503.

Killingsworth, Matthew A., and Daniel T. Gilbert. 2010. "A wandering mind is an unhappy mind." *Science* 330:932.

Kim, Young Shin, Eric Fombonne, Yun-Joo Koh, Soo-Jeong Kim, Keun-Ah Cheon, and Bennett L. Leventhal. 2014. "A comparison of DSM-IV pervasive developmental disorder and DSM-5 autism spectrum disorder prevalence in an epidemiologic sample." *J Am Acad Child Adolesc Psychiatry* 53 (5): 500–508.

Kim, Y. S., B. L. Leventhal, Y. J. Koh, E. Fombonne, E. Laska, E. C. Lim, K. A. Cheon, S. J. Kim, Y. K. Kim, H. Lee, D. H. Song, and R. R. Grinker. 2011. "Prevalence of autism spectrum disorders in a total population sample." *Am J Psychiatry* 168 (9): 904–12.

Kinnison, Michael T., and Andrew P. Hendry. 2001. "The pace of modern life II: From rates of contemporary microevolution to pattern and process." In *Microevolution rate, pattern, process*, edited by A. P. Hendry and M. T. Kinnison, 145–64. New York: Springer.

Kirkham, N. Z., J. A. Slemmer, and S. P. Johnson. 2002. "Visual statistical learning in infancy: Evidence for a domain general learning mechanism." *Cognition* 83 (2): B35–42.

Klei, Lambertus, Stephan J. Sanders, Michael T. Murtha, Vanessa Hus, Jennifer K. Lowe, A. Jeremy Willsey, Daniel Moreno-De-Luca, W. Yu Timothy, Eric Fombonne, and Daniel Geschwind. 2012. "Common genetic variants, acting additively, are a major source of risk for autism." *Mol Autism* 3 (1): 1–13.

Klin, Ami. 2000. "Attributing social meaning to ambiguous visual stimuli in higher-functioning autism and Asperger syndrome: The social attribution task." *J Child Psychol Psychiatry* 41 (7): 831–46.

Klin, Ami, David Pauls, Robert Schultz, and Fred Volkmar. 2005. "Three diagnostic approaches to Asperger syndrome: Implications for research." *J Autism Dev Disord* 35 (2): 221–34.

Knickmeyer, Rebecca C., Sylvain Gouttard, Chaeryon Kang, Dianne Evans, Kathy Wilber, J. Keith Smith, Robert M. Hamer, Weili Lin, Guido Gerig, and John H. Gilmore. 2008. "A structural MRI study of human brain development from birth to 2 years." *J Neurosci* 28 (47): 12176–82.

Kong, A., M. L. Frigge, G. Masson, S. Besenbacher, P. Sulem, G. Magnusson, S. A. Gudjonsson, A. Sigurdsson, A. Jonasdottir, W. S. Wong, G. Sigurdsson, G. B. Walters, S. Steinberg, H. Helgason, G. Thorleifsson, D. F. Gudbjartsson, A. Helgason, O. T. Magnusson, U. Thorsteinsdottir, and K. Stefansson. 2012. "Rate of *de novo* mutations and the importance of father's age to disease risk." *Nature* 488 (7412): 471–75.

Kopp, Svenny, and Christopher Gillberg. 1992. "Girls with social deficits and learning problems: Autism, atypical Asperger syndrome or a variant of these conditions." *Eur Child Adolesc Psychiatry* 1 (2): 89–99.

Kornmeier, J., R. Worner, A. Riedel, M. Bach, and L. Tebartz van Elst. 2014. "A different view on the checkerboard? Alterations in early and late visually evoked EEG potentials in Asperger observers." *PLoS One* 9 (3): e90993.

Krause, Johannes, Carles Lalueza-Fox, Ludovic Orlando, Wolfgang Enard, Richard E. Green, Hernán A. Burbano, Jean-Jacques Hublin, Catherine Hänni, Javier Fortea, and Marco De La Rasilla. 2007. "The derived *FOXP2* variant of modern humans was shared with Neandertals." *Curr Biol* 17 (21): 1908–12.

Kuhl, P. K. 2000. "A new view of language acquisition." *Proc Natl Acad Sci U S A* 97 (22): 11850–57.

———. 2004. "Early language acquisition: Cracking the speech code." *Nat Rev Neurosci* 5 (11): 831–43.

Kuhl, P. K., S. Coffey-Corina, D. Padden, and G. Dawson. 2005. "Links between social and linguistic processing of speech in preschool children with autism: Behavioral and electrophysiological measures." *Dev Sci* 8 (1): F1–F12.

Kuroda, Miho, Akio Wakabayashi, Tokio Uchiyama, Yuko Yoshida, Tomonori

Koyama, and Yoko Kamio. 2011. "Determining differences in social cognition between high-functioning autistic disorder and other pervasive developmental disorders using new advanced 'mind-reading' tasks." *Res Autism Spect Dis* 5 (1): 554–61.

Lai, M. C., M. V. Lombardo, and S. Baron-Cohen. 2014. "Autism." *Lancet* 383 (9920): 896–910.

Lai, Meng-Chuan, Michael V. Lombardo, Christine Ecker, Bhismadev Chakrabarti, John Suckling, Edward T. Bullmore, Francesca Happé, Declan G. M. Murphy, Simon Baron-Cohen, and MRC AIMS Consortium. 2015. "Neuro-anatomy of individual differences in language in adult males with autism." *Cereb Cortex* 25 (10): 3613–28.

Lai, M. C., M. V. Lombardo, G. Pasco, A. N. Ruigrok, S. J. Wheelwright, S. A. Sadek, B. Chakrabarti, Mrc Aims Consortium, and S. Baron-Cohen. 2011. "A behavioral comparison of male and female adults with high functioning autism spectrum conditions." *PLoS One* 6 (6): e20835.

Lainhart, J. E. 2015. "Brain imaging research in autism spectrum disorders: in search of neuropathology and health across the lifespan." *Curr Opin Psychiatry* 28 (2): 76–82.

Lancaster, J. B. 1985. "Evolutionary perspectives on sex differences in the higher primates." In *Gender and the life course*, edited by A. S. Rossi, 3–27. Hawthorne, N.Y.: Aldine de Gruyter.

Lange, N., B. G. Travers, E. D. Bigler, M. B. Prigge, A. L. Froehlich, J. A. Nielsen, A. N. Cariello, B. A. Zielinski, J. S. Anderson, P. T. Fletcher, A. A. Alexander, and J. E. Lainhart. 2015. "Longitudinal volumetric brain changes in autism spectrum disorder ages 6–35 years." *Autism Res* 8 (1): 82–93.

Langergraber, Kevin E., Kay Prüfer, Carolyn Rowney, Christophe Boesch, Catherine Crockford, Katie Fawcett, Eiji Inoue, Miho Inoue-Muruyama, John C. Mitani, and Martin N. Muller. 2012. "Generation times in wild chimpanzees and gorillas suggest earlier divergence times in great ape and human evolution." *Proc Natl Acad Sci U S A* 109 (39): 15716–21.

Larsen, F. W., and S. E. Mouridsen. 1997. "The outcome in children with childhood autism and Asperger syndrome originally diagnosed as psychotic: A 30-year follow-up study of subjects hospitalized as children." *Eur Child Adolesc Psychiatry* 6 (4): 181–90.

Lau, Yolanda C., Leighton B. N. Hinkley, Polina Bukshpun, Zoe A. Strominger, Mari L. J. Wakahiro, Simon Baron-Cohen, Carrie Allison, Bonnie Auyeung, Rita J. Jeremy, and Srikantan S. Nagarajan. 2013. "Autism traits in individuals with agenesis of the corpus callosum." *J Autism Dev Disord* 43 (5): 1106–18.

Lee, Richard B., and Irven DeVore, eds. 1968. *Man the hunter*. Chicago: Aldine.

Leekam, S. R., C. Nieto, S. J. Libby, L. Wing, and J. Gould. 2007. "Describing the sensory abnormalities of children and adults with autism." *J Autism Dev Disord* 37 (5): 894–910.

Leonard, Christiana M., Stephen Towler, Suzanne Welcome, Laura K. Halderman, Ron Otto, Mark A. Eckert, and Christine Chiarello. 2008. "Size matters: Cerebral volume influences sex differences in neuroanatomy." *Cereb Cortex* 18 (12): 2920–31.

Lesk, Arthur. 2014. *Introduction to bioinformatics.* 4th ed. New York: Oxford University Press.

Li, M., M. D. Fallin, A. Riley, R. Landa, S. O. Walker, M. Silverstein, D. Caruso, C. Pearson, S. Kiang, J. L. Dahm, X. Hong, G. Wang, M. C. Wang, B. Zuckerman, and X. Wang. 2016. "The association of maternal obesity and diabetes with autism and other developmental disabilities." *Pediatrics.* doi:10.1542/peds.2015-2206.

Lichtenstein, Paul, Eva Carlström, Maria Råstam, Christopher Gillberg, and Henrik Anckarsäter. 2010. "The genetics of autism spectrum disorders and related neuropsychiatric disorders in childhood." *Am J Psychiatry* 167 (11): 1357–63.

Lieberman, D. E. 2015. "Human locomotion and heat loss: An evolutionary perspective." *Compr Physiol* 5 (1): 99–117.

Lindell, A. K., and K. Hudry. 2013. "Atypicalities in cortical structure, handedness, and functional lateralization for language in autism spectrum disorders." *Neuropsychol Rev* 23 (3): 257–70.

Linnaeus, Carl von. 1735. *Systema saturae, sive regna tria naturae systematice proposita per classes, ordines, genera, & species.* Lugduni Batavorum [Leiden]: Johan Wilhelm de Groot.

Lippa, Richard A. 2010. "Sex differences in personality traits and gender-related occupational preferences across 53 nations: Testing evolutionary and social-environmental theories." *Arch Sex Behav* 39 (3): 619–36.

Liu, Meng-Jung, Wei-Lin Shih, and Le-Yin Ma. 2011. "Are children with Asperger syndrome creative in divergent thinking and feeling? A brief report." *Res Autism Spect Dis* 5 (1): 294–98.

Liu, Zhen, Xiao Li, Jun-Tao Zhang, Yi-Jun Cai, Tian-Lin Cheng, Cheng Cheng, Yan Wang, Chen-Chen Zhang, Yan-Hong Nie, and Zhi-Fang Chen. 2016. "Autism-like behaviours and germline transmission in transgenic monkeys overexpressing MeCP2." *Nature* 530:98–102.

Loh, Kep Kee, and Ryota Kanai. 2015. "How has the internet reshaped human cognition?" *Neuroscientist.* doi:10.1177/1073858415595005.

Lombardo, M. V., B. Chakrabarti, E. T. Bullmore, MRC AIMS Consortium, and S. Baron-Cohen. 2011. "Specialization of right temporo-parietal junction for mentalizing and its relation to social impairments in autism." *Neuroimage* 56 (3): 1832–38.

Lotter, Victor. 1978. "Childhood autism in Africa." *J Child Psychol Psychiatry* 19 (3): 231–44.

Luders, Eileen, Christian Gaser, Katherine L. Narr, and Arthur W. Toga. 2009. "Why sex matters: Brain size independent differences in gray matter distributions between men and women." *J Neurosci* 29 (45): 14265–70.

Lynch, Charles J., Lucina Q. Uddin, Kaustubh Supekar, Amirah Khouzam, Jennifer
    Phillips, and Vinod Menon. 2013. "Default mode network in childhood
    autism: posteromedial cortex heterogeneity and relationship with social
    deficits." *Biol Psychiat* 74 (3): 212–19.
Maher, Brendan. 2016. "Good gaming." *Nature* 531:568–72.
Mankoski, Raymond E., Martha Collins, Noah K. Ndosi, Ella H. Mgalla, Veronica V.
    Sarwatt, and Susan E. Folstein. 2006. "Etiologies of autism in a case-series
    from Tanzania." *J Autism Dev Disord* 36 (8): 1039–51.
Markram, Henry, Tania Rinaldi, and Kamila Markram. 2007. "The intense world
    syndrome—an alternative hypothesis for autism." *Front Neurosci* 1 (1):
    77–96.
Marlowe, Frank W. 2006. "Central place provisioning: The Hadza as an example." In
    *Feeding ecology in apes and other primates: Ecological, physical and behav-
    ioral aspects*, edited by G. Hohmann, N. M. Robbins and C. Boesch, 359–77.
    Cambridge: Cambridge University Press.
Mars, R. B., F. X. Neubert, M. P. Noonan, J. Sallet, I. Toni, and M. F. Rushworth.
    2012. "On the relationship between the 'default mode network' and the
    'social brain.'" *Front Hum Neurosci* 6:189.
Matson, Johnny L. and Alison M. Kozlowski. 2011. "The increasing prevalence of
    autism spectrum disorders." *Res Autism Spect Dis* 5 (1): 418–25.
Mattila, Marja-Leena, Marko Kielinen, Katja Jussila, Sirkka-Liisa Linna, Risto
    Bloigu, Hanna Ebeling, and Irma Moilanen. 2007. "An epidemiological and
    diagnostic study of Asperger syndrome according to four sets of diagnostic
    criteria." *J Am Acad Child Adolesc Psychiatry* 46 (5): 636–46.
Maximo, J. O., E. J. Cadena, and R. K. Kana. 2014. "The implications of brain con-
    nectivity in the neuropsychology of autism." *Neuropsychol Rev* 24 (1): 16–31.
McAlonan, G. M., C. Cheung, V. Cheung, N. Wong, J. Suckling, and S. E. Chua.
    2009. "Differential effects on white-matter systems in high-functioning
    autism and Asperger's syndrome." *Psychol Med* 39 (11): 1885–93.
McAlonan, G. M., J. Suckling, N. Wong, V. Cheung, N. Lienenkaemper, C. Cheung,
    and S. E. Chua. 2008. "Distinct patterns of grey matter abnormality in
    high-functioning autism and Asperger's syndrome." *J Child Psychol Psychi-
    atry* 49 (12): 1287–95.
McKelvey, J. Roger, Raymond Lambert, Laurent Mottron, and Michael I. Shevell.
    1995. "Right-hemisphere dysfunction in Asperger's syndrome." *J Child
    Neurol* 10 (4): 310–14.
McKinlay, Roger. 2016. "Technology: Use or lose our navigation skills." *Nature* 531
    (7596): 573–75.
McPartland, James, and Ami Klin. 2006. "Asperger's syndrome." *Adolesc Med Clin*
    17 (3): 771–88.
McPartland, James C., and F. R. Volkmar. 2013. "Asperger syndrome and its
    relationships to autism." In *Neuroscience of Autism Spectrum Disorders*,
    edited by J. D. Buxbaum and P. R. Hof, 55–67. New York: Academic Press.

McPherron, Shannon P., Zeresenay Alemseged, Curtis W. Marean, Jonathan G. Wynn, Denné Reed, Denis Geraads, René Bobe, and Hamdallah A. Béarat. 2010. "Evidence for stone-tool-assisted consumption of animal tissues before 3.39 million years ago at Dikika, Ethiopia." *Nature* 466 (7308): 857–60.

Meltzoff, Andrew N. 1999. "Origins of theory of mind, cognition and communication." *J Commun Disord* 32 (4): 251–69.

———. 2007. "'Like me': A foundation for social cognition." *Dev Sci* 10 (1): 126–34.

Meltzoff, A. N., and J. Decety. 2003. "What imitation tells us about social cognition: A rapprochement between developmental psychology and cognitive neuroscience." *Philos Trans R Soc Lond B Biol Sci* 358 (1431): 491–500.

Mendel, Gregor. 1866. "Versuche über pflanzenhybriden." *Verhandlungen des naturforschenden Vereines in Brunn* 4:3–47

Mesgarani, N., C. Cheung, K. Johnson, and E. F. Chang. 2014. "Phonetic feature encoding in human superior temporal gyrus." *Science* 343 (6174): 1006–10.

Meyer, M., M. Kircher, M. T. Gansauge, H. Li, F. Racimo, S. Mallick, J. G. Schraiber, F. Jay, K. Prufer, C. de Filippo, P. H. Sudmant, C. Alkan, Q. Fu, R. Do, N. Rohland, A. Tandon, M. Siebauer, R. E. Green, K. Bryc, A. W. Briggs, U. Stenzel, J. Dabney, J. Shendure, J. Kitzman, M. F. Hammer, M. V. Shunkov, A. P. Derevianko, N. Patterson, A. M. Andres, E. E. Eichler, M. Slatkin, D. Reich, J. Kelso, and S. Paabo. 2012. "A high-coverage genome sequence from an archaic Denisovan individual." *Science* 338 (6104): 222–26.

Miller, Gregory A., Laura D. Crocker, Jeffrey M. Spielberg, Zachary P. Infantolino, and Wendy Heller. 2013. "Issues in localization of brain function: the case of lateralized frontal cortex in cognition, emotion, and psychopathology." *Front Integr Neurosci* 7.

Mills, Kathryn L. 2014. "Effects of internet use on the adolescent brain: Despite popular claims, experimental evidence remains scarce." *Trends Cogn Sci* 18 (8): 385–87.

Moisse, Katie. 2016. "U.S. stats show autism rate reaching possible plateau." *Spectrum News*, March 31. https://spectrumnews.org/news/u-s-stats-show-autism-rate-reaching-possible-plateau.

Molnar-Szakacs, I. Uddin, and L. Q. Uddin. 2013. "Self-processing and the default mode network: Interactions with the mirror neuron system." *Front Hum Neurosci* 7:571.

Monastersky, R. 2015. "Anthropocene: The human age." *Nature* 519 (7542): 144–47.

Montgomery, Janine M., Adam W. McCrimmon, Vicki L. Schwean, and Donald H. Saklofske. 2010. "Emotional intelligence in Asperger syndrome: Implications of dissonance between intellect and affect." *Educ Train Autism Dev Disabil* 45 (4): 566–82.

Montgomery, Janine M., Brenda M. Stoesz, and Adam W. McCrimmon. 2013. "Emotional intelligence, theory of mind, and executive functions as predictors of social outcomes in young adults with Asperger syndrome." *Focus Autism Other Dev Disabl* 28 (1): 4–13.

Montiel-Nava, Cecilia, and Joaquín A. Peña. 2008. "Epidemiological findings of pervasive developmental disorders in a Venezuelan study." *Autism* 12 (2): 191–202.

Morgan, John T., Gursharan Chana, Ian Abramson, Katerina Semendeferi, Eric Courchesne, and Ian P. Everall. 2012. "Abnormal microglial-neuronal spatial organization in the dorsolateral prefrontal cortex in autism." *Brain Res* 1456:72–81.

Morgan, John T., Gursharan Chana, Carlos A. Pardo, Cristian Achim, Katerina Semendeferi, Jody Buckwalter, Eric Courchesne, and Ian P. Everall. 2010. "Microglial activation and increased microglial density observed in the dorsolateral prefrontal cortex in autism." *Biol Psychiat* 68 (4): 368–76.

Moseley, R. L., F. Pulvermuller, B. Mohr, M. V. Lombardo, S. Baron-Cohen, and Y. Shtyrov. 2014. "Brain routes for reading in adults with and without autism: EMEG evidence." *J Autism Dev Disord* 44 (1): 137–53.

Mouridsen, Svend Erik, Henrik Brønnum-Hansen, Bente Rich, and Torben Isager. 2008. "Mortality and causes of death in autism spectrum disorders: An update." *Autism* 12 (4): 403–14.

Mouridsen, Svend Erik, Bente Rich, and Torben Isager. 2002. "Body mass index in male and female children with infantile autism." *Autism* 6 (2): 197–205.

Mueller, S., D. Keeser, A. C. Samson, V. Kirsch, J. Blautzik, M. Grothe, O. Erat, M. Hegenloh, U. Coates, M. F. Reiser, K. Hennig-Fast, and T. Meindl. 2013. "Convergent findings of altered functional and structural brain connectivity in individuals with high functioning autism: A multimodal MRI study." *PLoS One* 8 (6): e67329.

Murphy, Clodagh M., Q. Deeley, E. M. Daly, C. Ecker, F. M. O'Brien, B. Hallahan, E. Loth, F. Toal, S. Reed, and S. Hales. 2012. "Anatomy and aging of the amygdala and hippocampus in autism spectrum disorder: An in vivo magnetic resonance imaging study of Asperger syndrome." *Autism Res* 5 (1): 3–12.

Muth, A., J. Honekopp, and C. M. Falter. 2014. "Visuo-spatial performance in autism: A meta-analysis." *J Autism Dev Disord* 44 (12): 3245–63.

Myles, Brenda, Taku Hagiwara, Winnie Dunn, Louann Rinner, Matthew Reese, Abby Huggins, and Stephanie Becker. 2004. "Sensory issues in children with Asperger syndrome." *Educ Train Autism Dev Disabil* 39 (4): 283–90.

Nielsen, Jared A., Brandon A. Zielinski, Michael A. Ferguson, Janet E. Lainhart, and Jeffrey S. Anderson. 2013. "An evaluation of the left-brain vs. right-brain hypothesis with resting state functional connectivity magnetic resonance imaging." *PloS One* 8 (8): e71275.

Nolte, John. 2009. *The human brain: An introduction to its functional anatomy.* 6th ed. Philadelphia: Mosby Elsevier.

Nomi, J. S. and L. Q. Uddin. 2015. "Developmental changes in large-scale network connectivity in autism." *Neuroimage Clin* 7:732–41.

Norbury, C. F., and A. Sparks. 2013. "Difference or disorder? Cultural issues in understanding neurodevelopmental disorders." *Dev Psychol* 49 (1): 45–58.

Nordahl, C. W., D. Dierker, I. Mostafavi, C. M. Schumann, S. M. Rivera, D. G. Amaral, and D. C. Van Essen. 2007. "Cortical folding abnormalities in autism revealed by surface-based morphometry." *J Neurosci* 27 (43): 11725–35.

Nordahl, C. W., R. Scholz, X. Yang, M. H. Buonocore, T. Simon, S. Rogers, and D. G. Amaral. 2012. "Increased rate of amygdala growth in children aged 2 to 4 years with autism spectrum disorders: A longitudinal study." *Arch Gen Psychiatry* 69 (1): 53–61.

Norris, Nola Grace. 2014. "A new perspective on thinking, memory and learning in gifted adults with Asperger syndrome: Five phenomenological case studies." PhD diss., University of Wollongong.

Nuttle, X., G. Giannuzzi, M. H. Duyzend, J. G. Schraiber, I. Narvaiza, P. H. Sudmant, O. Penn, G. Chiatante, M. Malig, J. Huddleston, C. Benner, F. Camponeschi, S. Ciofi-Baffoni, H. A. Stessman, M. C. Marchetto, L. Denman, L. Harshman, C. Baker, A. Raja, K. Penewit, N. Janke, W. J. Tang, M. Ventura, L. Banci, F. Antonacci, J. M. Akey, C. T. Amemiya, F. H. Gage, A. Reymond, and E. E. Eichler. 2016. "Emergence of a *Homo sapiens*-specific gene family and chromosome 16p11.2 CNV susceptibility." *Nature* 536 (7615): 205–9.

O'Callaghan, C., J. M. Shine, S. J. Lewis, J. R. Andrews-Hanna, and M. Irish. 2015. "Shaped by our thoughts—a new task to assess spontaneous cognition and its associated neural correlates in the default network." *Brain Cogn* 93:1–10.

Oksenberg, Nir, Laurie Stevison, Jeffrey D. Wall, and Nadav Ahituv. 2013. "Function and regulation of *AUTS2*, a gene implicated in autism and human evolution." *PLoS Genet* 9 (1): e1003221.

Ornitz, E. M. 1974. "The modulation of sensory input and motor output in autistic children." *J Autism Child Schizophr* 4 (3): 197–215.

Ouimet, Tia, Nicholas E. V. Foster, Ana Tryfon, and Krista L. Hyde. 2012. "Auditory-musical processing in autism spectrum disorders: A review of behavioral and brain imaging studies." *Ann NY Acad Sci* 1252 (1): 325–31.

Ouss, L., C. Saint-Georges, L. Robel, N. Bodeau, M. C. Laznik, G. C. Crespin, M. Chetouani, C. Bursztejn, B. Golse, R. Nabbout, I. Desguerre, and D. Cohen. 2014. "Infant's engagement and emotion as predictors of autism or intellectual disability in West syndrome." *Eur Child Adolesc Psychiatry* 23 (3): 143–49.

Ozonoff, S., S. Macari, G. S. Young, S. Goldring, M. Thompson, and S. J. Rogers. 2008. "Atypical object exploration at 12 months of age is associated with autism in a prospective sample." *Autism* 12 (5): 457–72.

Ozonoff, S., G. S. Young, S. Goldring, L. Greiss-Hess, A. M. Herrera, J. Steele, S. Macari, S. Hepburn, and S. J. Rogers. 2008. "Gross motor development, movement abnormalities, and early identification of autism." *J Autism Dev Disord* 38 (4): 644–56.

Padawer, R. 2014. "The recovered." *New York Times Magazine*, August 3, 2014, 20–27; 46–47.

Park, A. 2009. "Autism numbers are rising. The question is why?" *Time*. http://
    content.time.com/time/printout/0,8816,1948842,00.html.

Passingham, R. E. 1975. "Changes in the size and organisation of the brain in man
    and his ancestors." *Brain Behav Evol* 11 (2): 73–90.

Paula, Cristiane S., Sabrina H. Ribeiro, Eric Fombonne, and Marcos T. Mercadante.
    2011. "Brief report: Prevalence of pervasive developmental disorder in Bra-
    zil: A pilot study." *J Autism Dev Disord* 41 (12): 1738–42.

Paynter, Jessica, and Candida Peterson. 2010. "Language and ToM development in
    autism versus Asperger syndrome: Contrasting influences of syntactic
    versus lexical/semantic maturity." *Res Autism Spect Dis* 4 (3): 377–85.

Persico, Antonio M., and Valerio Napolioni. 2013. "Autism genetics." *Behav Brain Res*
    251:95–112.

Peters, Michael, and Christian Battista. 2008. "Applications of mental rotation fig-
    ures of the Shepard and Metzler type and description of a mental rotation
    stimulus library." *Brain Cogn* 66 (3): 260–64.

Peters, Michael, Bruno Laeng, Kerry Latham, Marla Jackson, Raghad Zaiyouna,
    and Chris Richardson. 1995. "A redrawn Vandenberg and Kuse mental
    rotations test-different versions and factors that affect performance."
    Brain Cogn 28 (1): 39–58.

Petković, Zorana Bujas, Vlatka Boričević Maršanić, Branka Divčić, and Nela Erce-
    gović. 2015. "Late diagnosis of Asperger syndrome in Croatia—a low-income
    country." *Psychiatria Danubina* 27 (4): 426–28.

Pichon, Swann, and Christian A. Kell. 2013. "Affective and sensorimotor components
    of emotional prosody generation." *J Neurosci* 33 (4): 1640–50.

Pinel, Philippe, and Stanislas Dehaene. 2010. "Beyond hemispheric dominance: Brain
    regions underlying the joint lateralization of language and arithmetic to
    the left hemisphere." *J Cogn Neurosci* 22 (1): 48–66.

Pinel, Philippe, Christophe Lalanne, Thomas Bourgeron, Fabien Fauchereau, Cyril
    Poupon, Eric Artiges, Denis Le Bihan, Ghislaine Dehaene-Lambertz, and
    Stanislas Dehaene. 2015. "Genetic and environmental influences on the
    visual word form and fusiform face areas." *Cereb Cortex* 25 (9): 2478–93.

Planche, Pascale, and Eric Lemonnier. 2012. "Children with high-functioning autism
    and Asperger's syndrome: Can we differentiate their cognitive profiles?"
    Res Autism Spect Dis 6 (2): 939–48.

Ploeger, Annemie, and Frietson Galis. 2011. "Evolutionary approaches to autism—an
    overview and integration." *Mcgill J Med* 13 (2): 38-43.

Plooij, F. X. 1984. *The behavioral development of free-living chimpanzee babies and
    infants*. Norwood, N.J.: Ablex.

Power, Robert A., Simon Kyaga, Rudolf Uher, James H. MacCabe, Niklas Långström,
    Mikael Landen, Peter McGuffin, Cathryn M. Lewis, Paul Lichtenstein, and
    Anna C. Svensson. 2013. "Fecundity of patients with schizophrenia, autism,
    bipolar disorder, depression, anorexia nervosa, or substance abuse vs their
    unaffected siblings." *JAMA Psychiatry* 70 (1): 22–30.

Premack, D., Woodruff, G. 1978. "Does the chimpanzee have a theory of mind?" *Behav Brain Sci* 1 (4): 515–26.

Preuss, Todd M. 2011. "The human brain: Rewired and running hot." *Ann N Y Acad Sci* 1225 (S1): E182–E191.

Provine, Robert R., Kurt A. Krosnowski, and Nicole W. Brocato. 2009. "Tearing: Breakthrough in human emotional signaling." *Evol Psychol* 7 (1): 52–56.

Prufer, K., F. Racimo, N. Patterson, F. Jay, S. Sankararaman, S. Sawyer, A. Heinze, G. Renaud, P. H. Sudmant, C. de Filippo, H. Li, S. Mallick, M. Dannemann, Q. Fu, M. Kircher, M. Kuhlwilm, M. Lachmann, M. Meyer, M. Ongyerth, M. Siebauer, C. Theunert, A. Tandon, P. Moorjani, J. Pickrell, J. C. Mullikin, S. H. Vohr, R. E. Green, I. Hellmann, P. L. Johnson, H. Blanche, H. Cann, J. O. Kitzman, J. Shendure, E. E. Eichler, E. S. Lein, T. E. Bakken, L. V. Golovanova, V. B. Doronichev, M. V. Shunkov, A. P. Derevianko, B. Viola, M. Slatkin, D. Reich, J. Kelso, and S. Paabo. 2014. "The complete genome sequence of a Neanderthal from the Altai Mountains." *Nature* 505 (7481): 43–49.

Pugliese, L., M. Catani, S. Ameis, F. Dell'Acqua, M. Thiebaut de Schotten, C. Murphy, D. Robertson, Q. Deeley, E. Daly, and D. G. Murphy. 2009. "The anatomy of extended limbic pathways in Asperger syndrome: A preliminary diffusion tensor imaging tractography study." *Neuroimage* 47 (2): 427–34.

Rane, P., D. Cochran, S. M. Hodge, C. Haselgrove, D. N. Kennedy, and J. A. Frazier. 2015. "Connectivity in autism: A review of MRI connectivity studies." *Harv Rev Psychiatry* 23 (4): 223–44.

Raznahan, Armin, Gregory L. Wallace, Ligia Antezana, Dede Greenstein, Rhoshel Lenroot, Audrey Thurm, Marta Gozzi, Sarah Spence, Alex Martin, and Susan E. Swedo. 2013. "Compared to what? Early brain overgrowth in autism and the perils of population norms." *Biol Psychiat* 74 (8): 563–75.

Reed, D. L., J. E. Light, J. M. Allen, and J. J. Kirchman. 2007. "Pair of lice lost or parasites regained: The evolutionary history of anthropoid primate lice." *BMC Biol* 5:7.

Reilly, Steven K., Jun Yin, Albert E. Ayoub, Deena Emera, Jing Leng, Justin Cotney, Richard Sarro, Pasko Rakic, and James P. Noonan. 2015. "Evolutionary changes in promoter and enhancer activity during human corticogenesis." *Science* 347 (6226): 1155–59.

Riquelme, Inmaculada, Samar M. Hatem, and Pedro Montoya. 2016. "Abnormal pressure pain, touch sensitivity, proprioception, and manual dexterity in children with autism spectrum disorders." *Neural Plasticity.* doi:10.1093/cercor/bhv125.

Rivet, Tessa Taylor, and Johnny L. Matson. 2011. "Gender differences in core symptomatology in autism spectrum disorders across the lifespan." *J Dev Phys Disabil* 23 (5): 399–420.

Robinson, John Elder. 2011. *Be different: My adventures with Asperger's and my advice for fellow Aspergians, misfits, families, and teachers.* New York: Broadway Paperbacks.

Roelfsema, Martine T., Rosa A. Hoekstra, Carrie Allison, Sally Wheelwright, Carol Brayne, Fiona E. Matthews, and Simon Baron-Cohen. 2012. "Are autism spectrum conditions more prevalent in an information-technology region? A school-based study of three regions in the Netherlands." *J Autism Dev Disord* 42 (5): 734–39.

Rolian, C., D. E. Lieberman, and B. Hallgrímsson. 2010. "The coevolution of human hands and feet." *Evolution* 64 (6): 1558–68.

Ronald, A., and R. A. Hoekstra. 2011. "Autism spectrum disorders and autistic traits: A decade of new twin studies." *Am J Med Genet B Neuropsychiatr Genet* 156B (3): 255–74.

Ronemus, Michael, Ivan Iossifov, Dan Levy, and Michael Wigler. 2014. "The role of *de novo* mutations in the genetics of autism spectrum disorders." *Nat Rev Genet* 15 (2): 133–41.

Rosenberg, K. R., and W. R. Trevathan. 2001. "Evolution of human birth." *Sci Am* 285 (5): 77–81.

———. 2002. "Birth, obstetrics and human evolution." *BJOG* 109 (11): 1199–206.

Ross, C. 2001. "Park or ride? Evolution of infant carrying in primates." *Int J Primatol* 22:749–71.

Rubenstein, John L. R. 2010. "Three hypotheses for developmental defects that may underlie some forms of autism spectrum disorder." *Curr Opin Neurol* 23 (2): 118–23.

Rudie, Jeffrey D., J. A. Brown, D. Beck-Pancer, L. M. Hernandez, E. L. Dennis, P. M. Thompson, S. Y. Bookheimer, and M. Dapretto. 2013. "Altered functional and structural brain network organization in autism." *Neuroimage Clin* 2:79–94.

Rueckl, Jay G., Pedro M. Paz-Alonso, Peter J. Molfese, Wen-Jui Kuo, Atira Bick, Stephen J. Frost, Roeland Hancock, Denise H. Wu, William Einar Mencl, and Jon Andoni Duñabeitia. 2015. "Universal brain signature of proficient reading: Evidence from four contrasting languages." *Proc Natl Acad Sci U S A* 112 (50): 15510–15.

Rueda, Pilar, Pablo Fernández-Berrocal, and Kimberly A. Schonert-Reichl. 2014. "Empathic abilities and theory of mind in adolescents with Asperger syndrome: Insights from the twenty-first century." *Rev J Autism Dev Disord* 1 (4): 327–43.

Ruigrok, Amber N. V., Gholamreza Salimi-Khorshidi, Meng-Chuan Lai, Simon Baron-Cohen, Michael V. Lombardo, Roger J. Tait, and John Suckling. 2014. "A meta-analysis of sex differences in human brain structure." *Neurosci Biobehav Rev* 39:34–50.

Ruthsatz, Joanne, and Kimberly Stephens. 2016. *The prodigy's cousin: The family link between autism and extraordinary talent.* New York: Penguin.

Rutter, Michael. 1978. "Language disorder and infantile autism." In *Autism: A reappraisal of concepts and treatment*, edited by M. Rutter and E. Schopler, 85–104. New York: Springer.

Rynkiewicz, Agnieszka, Björn Schuller, Erik Marchi, Stefano Piana, Antonio

Camurri, Amandine Lassalle, and Simon Baron-Cohen. 2016. "An investigation of the 'female camouflage effect' in autism using a computerized ADOS-2 and a test of sex/gender differences." *Mol Autism* 7 (1): 1–8.

Rysstad, Anne Langseth, and Arve Vorland Pedersen. 2015. "Brief Report: Non-right-handedness within the autism spectrum disorder." *J Autism Dev Disord*:1–8.

Saban-Bezalel, R., and N. Mashal. 2015. "Hemispheric processing of idioms and irony in adults with and without pervasive developmental disorder." *J Autism Dev Disord* 45 (11): 3496–508.

Sainsbury, Clare. 2004. *Martian in the playground: Understanding the schoolchild with Asperger's syndrome.* Bristol, UK: Antony Rowe.

Saint-Georges, C., M. Chetouani, R. Cassel, F. Apicella, A. Mahdhaoui, F. Muratori, M. C. Laznik, and D. Cohen. 2013. "Motherese in interaction: At the crossroad of emotion and cognition? (A systematic review)." *PLoS One* 8 (10): e78103.

Sakai, Tomoko, Satoshi Hirata, Kohki Fuwa, Keiko Sugama, Kiyo Kusunoki, Haruyuki Makishima, Tatsuya Eguchi, Shigehito Yamada, Naomichi Ogihara, and Hideko Takeshita. 2012. "Fetal brain development in chimpanzees versus humans." *Curr Biol* 22 (18): R791–R792.

Salyakina, D., D. Q. Ma, J. M. Jaworski, I. Konidari, P. L. Whitehead, R. Henson, D. Martinez, J. L. Robinson, S. Sacharow, and H. H. Wright. 2010. "Variants in several genomic regions associated with Asperger Disorder." *Autism Res* 3 (6): 303–10.

Sandin, Sven, Christina M. Hultman, Alexander Kolevzon, Raz Gross, James H. MacCabe, and Abraham Reichenberg. 2012. "Advancing maternal age is associated with increasing risk for autism: A review and meta-analysis." *J Am Acad Child Adolesc Psychiatry* 51 (5): 477–86.

Sandin, S., P. Lichtenstein, R. Kuja-Halkola, H. Larsson, C. M. Hultman, and A. Reichenberg. 2014. "The familial risk of autism." *JAMA* 311 (17): 1770–77.

Sauer, Sebastian, Harald Walach, Stefan Schmidt, Thilo Hinterberger, Siobhan Lynch, Arndt Büssing, and Niko Kohls. 2013. "Assessment of mindfulness: Review on state of the art." *Mindfulness* 4 (1): 3–17.

Schipul, Sarah E., Timothy A. Keller, and Marcel Adam Just. 2011. "Inter-regional brain communication and its disturbance in autism." *Front Syst Neurosci* 5.

Schmidt, K. L. and Cohn, J. F. 2001. "Human facial expressions as adaptations: Evolutionary questions in facial expression research." *Yearb Phys Anthropol* 44:3–24.

Schmitt, David P. 2015. "The evolution of culturally-variable sex differences: Men and women are not always different, but when they are . . . it appears *not* to result from patriarchy or sex role socialization." In *The evolution of sexuality,* edited by V. A. Weekes-Shackelford and T. K. Shackelford, 221–56. New York: Springer.

Schoenmakers, Wilfred, and Geert Duysters, 2010. "The technological origins of radical inventions." *Res Policy* 39 (8): 1051–59.

Schore, A. N. 2014. "The right brain is dominant in psychotherapy." *Psychotherapy* 51 (3): 388–97.

Schultz, Adolph H. 1941. "The relative size of the cranial capacity in primates." *Am J Phys Anthropol* 28 (3): 273–87.

Schumacher, Johannes, Per Hoffmann, Christine Schmäl, Gerd Schulte-Körne, and Markus M. Nöthen. 2007. "Genetics of dyslexia: The evolving landscape." *J Med Genet* 44 (5): 289–97.

Schumann, C. M., C. S. Bloss, C. C. Barnes, G. M. Wideman, R. A. Carper, N. Akshoomoff, K. Pierce, D. Hagler, N. Schork, C. Lord, and E. Courchesne. 2010. "Longitudinal magnetic resonance imaging study of cortical development through early childhood in autism." *J Neurosci* 30 (12): 4419–27.

Schwarz, E., P. C. Guest, H. Rahmoune, L. Wang, Y. Levin, E. Ingudomnukul, L. Ruta, L. Kent, M. Spain, S. Baron-Cohen, and S. Bahn. 2011. "Sex-specific serum biomarker patterns in adults with Asperger's syndrome." *Mol Psychiatry* 16 (12): 1213–20.

Sebat, Jonathan, B. Lakshmi, Dheeraj Malhotra, Jennifer Troge, Christa Lese-Martin, Tom Walsh, Boris Yamrom, Seungtai Yoon, Alex Krasnitz, and Jude Kendall. 2007. "Strong association of *de novo* copy number mutations with autism." *Science* 316 (5823): 445–49.

Seeley, William W., Vinod Menon, Alan F. Schatzberg, Jennifer Keller, Gary H. Glover, Heather Kenna, Allan L. Reiss, and Michael D. Greicius. 2007. "Dissociable intrinsic connectivity networks for salience processing and executive control." *J Neurosci* 27 (9): 2349–56.

Semendeferi, Katerina, Kate Teffer, Dan P. Buxhoeveden, Min S. Park, Sebastian Bludau, Katrin Amunts, Katie Travis, and Joseph Buckwalter. 2011. "Spatial organization of neurons in the frontal pole sets humans apart from great apes." *Cereb Cortex* 21 (7): 1485–97.

Sestieri, C., M. Corbetta, S. Spadone, G. L. Romani, and G. L. Shulman. 2014. "Domain-general signals in the cingulo-opercular network for visuospatial attention and episodic memory." *J Cogn Neurosci* 26 (3): 551–68.

Shepard, R. N., and J. Metzler. 1971. "Mental rotation of three-dimensional objects." *Science* 171 (3972): 701–3.

Shore, Stephen M. 2001. *Beyond the wall: Personal experiences with autism and Asperger syndrome*. Shawnee Mission, Kans.: Autism Asperger Publishing.

Silberman, S. 2001. "The geek syndrome." *Wired*, 9, 12. http://www.wired.com/wired/archive/9.12/aspergers. html.

———. 2015. *Neurotribes: The legacy of autism and the future of neurodiversity*. New York: Penguin.

Silk, Timothy J., Nicole Rinehart, John L. Bradshaw, Bruce Tonge, Gary Egan, Michael W. O'Boyle, and Ross Cunnington. 2006. "Visuospatial processing and the function of prefrontal-parietal networks in autism spectrum disorders: A functional MRI study." *Am J Psychiatry* 163 (8): 1440–43.

Simone, Rudy. 2010. *Aspergirls*. Philadelphia: Jessica Kingsley.

Singer, Judy. 1999. "Why can't you be normal for once in your life? From a problem with no name to the emergence of a new category of difference." *Disability Discourse*, 59–70.

Singer, Tania, Hugo D. Critchley, and Kerstin Preuschoff. 2009. "A common role of insula in feelings, empathy and uncertainty." *Trends Cogn Sci* 13 (8): 334–40.

Singh, Jasjit, and Lee Fleming. 2010. "Lone inventors as sources of breakthroughs: Myth or reality?" *Manage Sci* 56 (1): 41–56.

Slaughter, Virginia. 2015. "Theory of mind in infants and young children: A review." *Aust Psychol* 50 (3): 169–72.

Slocum, Sally. 1975. "Woman the gatherer: Male bias in anthropology." In *Toward an anthropology of women*, edited by Rayna Reiter, 36–50. New York: Monthly Review Press.

Small, Gary W., Teena D. Moody, Prabha Siddarth, and Susan Y. Bookheimer. 2009. "Your brain on Google: Patterns of cerebral activation during internet searching." *Am J Geriatr Psychiatry* 17 (2): 116–26.

Small, Meredith F. 1998. *Our babies, ourselves: How biology and culture shape the way we parent*. New York: Anchor Books.

Smith, Aaron C. T. 2008. "The neuroscience of spiritual experience in organizations." *Journal of Management, Spirituality & Religion* 5 (1): 3–28.

Smith, Heather F. 2011. "The role of genetic drift in shaping modern human cranial evolution: A test using microevolutionary modeling." *Int J Evol Biol* 2011:1–11.

Smith, Stephen M., Peter T. Fox, Karla L. Miller, David C. Glahn, P. Mickle Fox, Clare E. Mackay, Nicola Filippini, Kate E. Watkins, Roberto Toro, and Angela R. Laird. 2009. "Correspondence of the brain's functional architecture during activation and rest." *Proc Natl Acad Sci U S A* 106 (31): 13040–45.

Smuts, Barbara. 1992. "Male aggression against women." *Hum Nature* 3 (1): 1–44.

Sobanski, E., A. Marcus, K. Hennighausen, J. Hebebrand, and M. H. Schmidt. 1999. "Further evidence for a low body weight in male children and adolescents with Asperger's disorder." *Eur Child Adolesc Psychiatry* 8 (4): 312–14.

Soden, Brooke, Micaela E. Christopher, Jacqueline Hulslander, Richard K. Olson, Laurie Cutting, Janice M. Keenan, Lee A. Thompson, Sally J. Wadsworth, Erik G. Willcutt, and Stephen A. Petrill. 2015. "Longitudinal stability in reading comprehension is largely heritable from grades 1 to 6." *PloS One* 10 (1): e0113807.

Sowell, Elizabeth R., Bradley S. Peterson, Eric Kan, Roger P. Woods, June Yoshii, Ravi Bansal, Dongrong Xu, Hongtu Zhu, Paul M. Thompson, and Arthur W. Toga. 2007. "Sex differences in cortical thickness mapped in 176 healthy individuals between 7 and 87 years of age." *Cereb Cortex* 17 (7): 1550–60.

Spek, A. A., E. M. Scholte, and I. A. Van Berckelaer-Onnes. 2010. "Theory of mind in adults with HFA and Asperger syndrome." *J Autism Dev Disord* 40 (3): 280–89.

Spek, Annelies A., Nadia C. van Ham, and Ivan Nyklíček. 2013. "Mindfulness-based therapy in adults with an autism spectrum disorder: A randomized controlled trial." *Res Dev Disabil* 34 (1): 246–53.

Spek, Annelies A., and E. Velderman. 2013. "Examining the relationship between Autism spectrum disorders and technical professions in high functioning adults." *Res Autism Spect Dis* 7 (5): 606–12.

Sperry, Roger W. 1968. "Hemisphere deconnection and unity in conscious awareness." *Am Psychol* 23 (10): 723–33.

Spikins, Penny. 2013. *The stone age origins of autism*. InTech. doi:10.5772/53883.

Spreng, R. N. 2012. "The fallacy of a 'task-negative' network." *Front Psychol* 3:145. doi:10.3389/fpsyg.2012.00145.

Squire, Larry, Darwin Berg, Floyd Bloom, Sascha du Lac, Anirvan Ghosh, and Nicholas Spitzer. 2008. *Fundamental neuroscience*. Boston: Academic Press.

Steinberg, Paul. 2012. "Asperger's history of overdiagnosis." *New York Times*, January 31, 2012. http://www.nytimes.com/2012/02/01/opinion/aspergers-history-of-over-diagnosis.html?_r=0/.

Stoet, Gijsbert, and David C. Geary. 2015. "Sex differences in academic achievement are not related to political, economic, or social equality." *Intelligence* 48:137–51.

Sun, Xiang, Carrie Allison, Fiona E. Matthews, Zhixiang Zhang, Bonnie Auyeung, Simon Baron-Cohen, and Carol Brayne. 2015. "Exploring the underdiagnosis and prevalence of autism spectrum conditions in Beijing." *Autism Res* 8 (3): 250–60.

Suren, P., C. Roth, M. Bresnahan, M. Haugen, M. Hornig, D. Hirtz, K. K. Lie, W. I. Lipkin, P. Magnus, T. Reichborn-Kjennerud, S. Schjolberg, G. Davey Smith, A. S. Oyen, E. Susser, and C. Stoltenberg. 2013. "Association between maternal use of folic acid supplements and risk of autism spectrum disorders in children." *JAMA* 309 (6): 570–77.

Szatmari, P., S. Bryson, E. Duku, L. Vaccarella, L. Zwaigenbaum, T. Bennett, and M. H. Boyle. 2009. "Similar developmental trajectories in autism and Asperger syndrome: From early childhood to adolescence." *J Child Psychol Psychiatry* 50 (12): 1459–67.

Szatmari, Peter, Xiao-Qing Liu, Jeremy Goldberg, Lonnie Zwaigenbaum, Andrew D. Paterson, Marc Woodbury-Smith, Stelios Georgiades, Eric Duku, and Ann Thompson. 2012. "Sex differences in repetitive stereotyped behaviors in autism: Implications for genetic liability." *Am J Med Genet B Neuropsychiatr Genet* 159 (1): 5–12.

Takahashi, D. Y., A. R. Fenley, Y. Teramoto, D. Z. Narayanan, J. I. Borjon, P. Holmes, and A. A. Ghazanfar. 2015. "Language development. The developmental dynamics of marmoset monkey vocal production." *Science* 349 (6249): 734–38.

Tamir, D. I., A. B. Bricker, D. Dodell-Feder, and J. P. Mitchell. 2015. "Reading fiction and reading minds: The role of simulation in the default network." *Soc Cogn Affect Neurosci.* doi:10.1093/scan/nsv114.

Tammet, Daniel. 2006. *Born on a blue day: A memoir of Asperger's and an extraordinary mind.* London: Hodder & Stoughton.

Tanguay, P. E. 2011. "Autism in DSM-5." *Am J Psychiatry* 168 (11): 1142–44.

Tanidir, Canan, and Nahit M. Mukaddes. 2014. "Referral pattern and special interests in children and adolescents with Asperger syndrome: A Turkish referred sample." *Autism* 18 (2): 178–84.

Taylor, Brent, Elizabeth Miller, C. Paddy Farrington, Maria-Christina Petropoulos, Isabelle Favot-Mayaud, Jun Li, and Paulie A. Waight. 1999. "Autism and measles, mumps, and rubella vaccine: No epidemiological evidence for a causal association." *Lancet* 353 (9169): 2026–29.

Teffer, K., and K. Semendeferi. 2012. "Human prefrontal cortex: Evolution, development, and pathology." In *Evolution of the primate brain: From neuron to behavior*, edited by M. A. Hofman and D. Falk, 191–218. London: Elsevier.

Teulier, C., K. Lee do, and B. D. Ulrich. 2015. "Early gait development in human infants: Plasticity and clinical applications." *Dev Psychobiol* 57 (4): 447–58.

Thurm, B. E., E. S. Pereira, C. C. Fonseca, M. J. S. Cagno, and E. F. Gama. 2011. "Neuroanatomical aspects of the body awareness." *J Morphol* 28 (4): 296–99.

Tick, Beata, Patrick Bolton, Francesca Happé, Michael Rutter, and Frühling Rijsdijk. 2016. "Heritability of autism spectrum disorders: a meta-analysis of twin studies." *J Child Psychol Psychiatry* 57(5): 585–95.

Toma, Claudio, Amaia Hervas, Barbara Torrico, Noemí Balmaña, Marta Salgado, Marta Maristany, Elisabet Vilella, Rafael Martínez-Leal, Ma Inmaculada Planelles, and Ivon Cusco. 2013. "Analysis of two language-related genes in autism: A case-control association study of *FOXP2* and *CNTNAP2*." *Psychiatr Genet* 23 (2): 82–85.

Tomasi, D., and N. D. Volkow. 2011. "Laterality patterns of brain functional connectivity: Gender effects." *Cereb Cortex* 22 (6): 1455–62.

Tottenham, N., M. E. Hertzig, K. Gillespie-Lynch, T. Gilhooly, A. J. Millner, and B. J. Casey. 2014. "Elevated amygdala response to faces and gaze aversion in autism spectrum disorder." *Soc Cogn Affect Neurosci* 9 (1): 106–17.

Trevathan, W., and K. Rosenberg, eds. 2016. *Costly and cute: How helpless newborns made us human.* Santa Fe, N.M.: SAR Press; Albuquerque: University of New Mexico Press.

Trivers, Robert. 1972. "Parental investment and sexual selection." In *Sexual selection and the descent of man*, edited by B. Campbell, 136–79. New York: Aldine de Gruyter.

Tsai, L. Y. 2013. "Asperger's disorder will be back." *J Autism Dev Disord* 43 (12): 2914–42.

Tsai, L. Y., and M. Ghaziuddin. 2014. "DSM-5 ASD moves forward into the past." *J Autism Dev Disord* 44 (2): 321–30.

Tseng, Yi-Li, Han Hsuan Yang, Alexander N. Savostyanov, Vincent S. C. Chien, and
    Michelle Liou. 2015. "Voluntary attention in Asperger's syndrome: Brain
    electrical oscillation and phase-synchronization during facial emotion
    recognition." *Res Autism Spect Dis* 13:32–51.
Uddin, Lucina Q., Kaustubh Supekar, Charles J. Lynch, Amirah Khouzam, Jennifer
    Phillips, Carl Feinstein, Srikanth Ryali, and Vinod Menon. 2013. "Salience
    network–based classification and prediction of symptom severity in chil-
    dren with autism." *JAMA Psychiatry* 70 (8): 869–79.
Uher, R. 2009. "The role of genetic variation in the causation of mental illness: An
    evolution-informed framework." *Mol Psychiatry* 14 (12): 1072–82.
Ulrich, Martin, Johannes Keller, Klaus Hoenig, Christiane Waller, and Georg Grön.
    2014. "Neural correlates of experimentally induced flow experiences."
    Neuroimage 86:194–202.
United Nations Educational, Scientific and Cultural Organization Institute for Statis-
    tics. 2014. "Adult and youth literacy." Fact Sheet No. 29. September. http://
    www.uis.unesco.org/literacy/Documents/fs-29-2014-literacy-en.pdf.
Vadée-Le-Brun, Yoram, Jonathan Rouzaud-Cornabas, and Guillaume Beslon. 2015.
    "Epigenetic inheritance speeds up evolution of artificial organisms." Euro-
    pean Conference on Artificial Life. https://mitpress.mit.edu/sites/default/
    files/titles/content/eca12015/978-0-262-33027-5-ch078.pdf.
van den Bos, Ruud. 2015. "Sex matters, as do individual differences . . ." *Trends Neu-
    rosci* 38 (7): 401–2.
van der Aa, Christine, Monique M. H. Pollmann, Aske Plaat, and Rutger Jan van
    der Gaag. 2016. "Computer-mediated communication in adults with
    high-functioning autism spectrum disorders and controls." *Res Autism
    Spect Dis* 23:15–27.
van der Knaap, Lisette J. and Ineke J. M. van der Ham. 2011. "How does the corpus
    callosum mediate interhemispheric transfer? A review." *Behav Brain Res*
    223 (1): 211–21.
Vandermassen, Griet. 2011. "Evolution and rape: A feminist Darwinian perspective."
    Sex Roles 64 (9–10): 732–47.
Vara, Vauhini. 2016. "Brain trust." *FastCompany.com*, September.
    https://www.fastcompany.com/40405949/this-pharrell-assisted-
    videos-use-of-split-screen-will-blow-your-mind.
Vernot, B., S. Tucci, J. Kelso, J. G. Schraiber, A. B. Wolf, R. M. Gittelman,
    M. Dannemann, S. Grote, R. C. McCoy, H. Norton, L. B. Scheinfeldt, D. A.
    Merriwether, G. Koki, J. S. Friedlaender, J. Wakefield, S. Paabo, and J. M.
    Akey. 2016. "Excavating Neandertal and Denisovan DNA from the genomes
    of Melanesian individuals." *Science* 352 (6282): 235–39.
Vidal, C. N., R. Nicolson, T. J. DeVito, K. M. Hayashi, J. A. Geaga, D. J. Drost, P. C.
    Williamson, N. Rajakumar, Y. Sui, R. A. Dutton, A. W. Toga, and P. M.
    Thompson. 2006. "Mapping corpus callosum deficits in autism: an index of
    aberrant cortical connectivity." *Biol Psychiat* 60 (3): 218–25.

Volkmar, F. R., A. Klin, R. T. Schultz, E. Rubin, and R. Bronen. 2000. "Asperger's disorder." *Am J Psychiatry* 157 (2): 262–67.

Wall-Scheffler, C. M., K. Geiger, and K. L. Steudel-Numbers. 2007. "Infant carrying: The role of increased locomotory costs in early tool development." *Am J Phys Anthropol* 133 (2): 841–46.

Walter, Henrik. 2012. "Social cognitive neuroscience of empathy: Concepts, circuits, and genes." *Emot Rev* 4 (1): 9–17.

Warrier, Varun, Bhismadev Chakrabarti, Laura Murphy, Allen Chan, Ian Craig, Uma Mallya, Silvia Lakatošová, Karola Rehnstrom, Leena Peltonen, and Sally Wheelwright. 2015. "A pooled genome-wide association study of Asperger syndrome." *PloS One* 10 (7): e0131202.

Warrier, Varun, Simon Baron-Cohen, and Bhismadev Chakrabarti. 2013. "Genetic variation in GABRB3 is associated with Asperger syndrome and multiple endophenotypes relevant to autism." *Mol Autism* 4 (1): 48.

Washburn, S. L. 1960. "Tools and human evolution." *Sci Am* 203:63–75.

Watanabe, Hiroyuki, Sylvia Fitting, Muhammad Z. Hussain, Olga Kononenko, Anna Iatsyshyna, Takashi Yoshitake, Jan Kehr, Kanar Alkass, Henrik Druid, and Henrik Wadensten. 2015. "Asymmetry of the endogenous opioid system in the human anterior cingulate: A putative molecular basis for lateralization of emotions and pain." *Cereb Cortex* 25 (1): 97–108.

Watson, J. D., and F. H. Crick. 1953. "Genetical implications of the structure of deoxyribonucleic acid." *Nature* 171 (4361): 964–67.

Weimer, A. K., A. M. Schatz, A. Lincoln, A. O. Ballantyne, and D. A. Trauner. 2001. "'Motor' impairment in Asperger syndrome: Evidence for a deficit in proprioception." *J Dev Behav Pediatr* 22 (2): 92–101.

Welchew, D. E., C. Ashwin, K. Berkouk, R. Salvador, J. Suckling, S. Baron-Cohen, and E. Bullmore. 2005. "Functional disconnectivity of the medial temporal lobe in Asperger's syndrome." *Biol Psychiat* 57 (9): 991–8.

Werling, Donna M., and Daniel H. Geschwind. 2013. "Sex differences in autism spectrum disorders." *Curr Opin Neurol* 26 (2): 146–153.

Wermke, K., D. Leising, and A. Stellzig-Eisenhauer. 2007. "Relation of melody complexity in infants' cries to language outcome in the second year of life: A longitudinal study." *Clin Linguist Phon* 21 (11–12): 961–73.

White, Keith D. 2014. Abnormal vestibulo-ocular reflexes in autism: A potential endophenotype. DTIC document. http://oai.dtic.mil/oai/oai?verb=getRecord&metadataPrefix=html&identifier=ADA612857.

Whyte, Elisabeth M., Marlene Behrmann, Nancy J. Minshew, Natalie V. Garcia, and K. Suzanne Scherf. 2015. "Animal, but not human, faces engage the distributed face network in adolescents with autism." *Dev Sci* 19 (2): 306-317.

Willey, Angela, Banu Subramaniam, Jennifer A. Hamilton, and Jane Couperus. 2015. "The mating life of geeks: Love, neuroscience, and the new autistic subject." *Signs* 40 (2): 369–91.

Willey, Liane Holliday. 1999. *Pretending to be normal: Living with Asperger's syndrome*. Philadelphia: Jessica Kingsley.

———, ed. 2003. *Asperger syndrome in adolescence: Living with the ups, the downs, and things in between*. Philadelphia: Jessica Kingsley.

———. 2012. *Safety skills for Asperger women: How to save a perfectly good female life*. Philadelphia: Jessica Kingsley.

Williams, D. L., V. L. Cherkassky, R. A. Mason, T. A. Keller, N. J. Minshew, and M. A. Just. 2013. "Brain function differences in language processing in children and adults with autism." *Autism Res* 6 (4): 288–302.

Williams, Jo, Fiona Scott, Carol Stott, Carrie Allison, Patrick Bolton, Simon Baron-Cohen, and Carol Brayne. 2005. "The CAST (Childhood Asperger Syndrome Test) Test accuracy." *Autism* 9 (1): 45–68.

Wolf, Maryanne, and Mirit Barzillai. 2009. "The importance of deep reading." *Educ Leadersh* 66 (6): 32–37.

Wood, Andrew R., Tonu Esko, Jian Yang, Sailaja Vedantam, Tune H. Pers, Stefan Gustafsson, Audrey Y. Chu, Karol Estrada, Jian'an Luan, and Zoltán Kutalik. 2014. "Defining the role of common variation in the genomic and biological architecture of adult human height." *Nat Genet* 46 (11): 1173–86.

World Health Organization. 1992. *The ICD-10 classification of mental and behavioural disorders: Clinical descriptions and diagnostic guidelines*. Geneva: World Health Organization.

Wright, Jessica. 2015. "'CRISPR' way to cut genes speeds advances in autism." *Spectrum News*, December 14. https://spectrumnews.org/news/crispr-way-to-cut-genes-speeds-advances-in-autism.

Xu, Linda. 2014. "Humans computers and everything in between: Towards synthetic telepathy." *Harvard Sci Rev*, May 1, 2014. https://harvardsciencereview.com/2014/05/01/synthetic-telepathy/.

Yu, Cassie, and Bryan H. King. 2016. "Focus on Autism and Related Conditions." *FOCUS* 14 (1): 1–8.

Zablotsky, B., L. I. Black, M. J. Maenner, L. A. Schieve, and S. J. Blumberg. 2015. "Estimated prevalence of autism and other developmental disabilities following questionnaire changes in the 2014 National Health Interview Survey." *Natl Health Stat Report* (87): 1–21.

Zago, Laure, Laurent Petit, Emmanuel Mellet, Gaël Jobard, Fabrice Crivello, Marc Joliot, Bernard Mazoyer, and Nathalie Tzourio-Mazoyer. 2015. "The association between hemispheric specialization for language production and for spatial attention depends on left-hand preference strength." *Neuropsychologia* 93:394–406.

Zapf, A. C., L. A. Glindemann, K. Vogeley, and C. M. Falter. 2015. "Sex differences in mental rotation and how they add to the understanding of autism." *PLoS One* 10 (4): e0124628.

Zeliadt, Nicholette. 2016. "Where the vocabulary of autism is failing." *Spectrum*

News, April 1. http://www.theatlantic.com/hcalth/archive/2016/04/
the-language-of-autism/476223.

Zhang, Jie, John J. Wheeler, and Dean Richey. 2006. "Cultural validity in assessment
instruments for children with autism from a Chinese cultural perspective."
Int J Spec Educ 21 (1): 109–14.

Zihlman, Adrienne. 1981. "Women as shapers of the human adaptation." In *Woman
the gatherer*, edited by F. Dahlbert, 75–120. New Haven, Conn.: Yale University Press.

Zihlman, Adrienne L., and Carol E. Underwood. Forthcoming. *Comparative ape
anatomy and evolution*.

Page numbers in italic text indicate illustrations.